# SpringerBriefs in Space Development

*Series Editor*

Joseph N. Pelton

For further volumes:
http://www.springer.com/series/10058

Su-Yin Tan

# Meteorological Satellite Systems

Su-Yin Tan
Dept. of Geography and Environmental Mgt
University of Waterloo
Waterloo
Ontario
Canada

ISSN 2191-8171          ISSN 2191-818X   (electronic)
ISBN 978-1-4614-9419-5          ISBN 978-1-4614-9420-1   (eBook)
DOI 10.1007/978-1-4614-9420-1
Springer New York Heidelberg Dordrecht London

Library of Congress Control Number: 2013951777

Printed on acid-free paper

Springer is part of Springer Science+Business Media (www.springer.com)

*To my family*

# Acknowledgements

I am especially grateful to the series editor, Dr. Joseph N. Pelton, for his support, encouragement, and trust during the process of writing this book. His comments and enthusiasm were invaluable to improving the quality of the final manuscript. I would also like to express my gratitude to the publisher, Springer, for accepting this book as part of the series "SpringerBriefs in Space Development." In particular I would like to thank Maury Solomon and Nora Rawn for their editorial and production guidance.

My appreciation goes to my friends and colleagues at the Department of Geography and Environmental Management at the University of Waterloo. Thanks are due to Andre Roy, Peter Deadman, Jean Andrey, Paul Parker, and Ian McKenzie for their support. I would like to especially acknowledge Dongrong Li, who worked as my Research Assistant, as well as my graduate students at the Applied Geomatics Research Laboratory (AGRL).

I would like to thank the International Space University (ISU), which made it possible to write this book. As one of the most unique educational institutions in the world that promotes 3 "I"s (interdisciplinary, international, and intercultural) learning, I am grateful to be part of this visionary and innovative organization. Special thanks are due in particular to Angie Bukley, Walter Peeters, and the ISU Academic Council.

Finally, I would like to thank my mother and father, Wai-Koon and Geok Yong Tan, who have given me much encouragement and support over the years. This book is dedicated to Tomohisa Oki, my husband and 'best friend,' for his kindness, love, and dedication. I am deeply indebted to his unyielding support, especially when it mattered most.

# Acknowledgements

This Springer book is published in collaboration with the International Space University. At its central campus in Strasbourg, France, and at various locations around the world, the ISU provides graduate-level training to the future leaders of the global space community. The university offers a two-month Space Studies Program, a five-week Southern Hemisphere Program, a one-year Executive MBA and a one-year Masters program related to space science, space engineering, systems engineering, space policy and law, business and management, and space and society.

These programs give international graduate students and young space professionals the opportunity to learn while solving complex problems in an intercultural environment. Since its founding in 1987, the International Space University has graduated more than 3,000 students from 100 countries, creating an international network of professionals and leaders. ISU faculty and lecturers from around the world have published hundreds of books and articles on space exploration, applications, science and development.

# Contents

# About the Author

**Dr. Su-Yin Tan** is a senior lecturer at the University of Waterloo (Canada) and director of the Applied Geomatics Research Laboratory (AGRL). She teaches courses on geographic information systems, remote sensing, and spatial data analysis. She is a distinguished Gates Scholar and received her Ph.D. degree from the University of Cambridge (UK), two Masters degrees from Oxford University (UK) and Boston University (USA), and a BSc (Env) from the University of Guelph (Canada).

She has also chaired the Department of Space Applications and regularly serves as a visiting lecturer at the International Space University's Space Studies Program. Dr. Tan has an interdisciplinary background in the environmental sciences and spatial data analysis methodologies in a range of application areas, such as climatology, ecosystem modeling, and remote sensing. She has built a diverse record of research experience in North America, Australia, Asia, South America, and Europe. Although born in Canada, she was raised in Papua New Guinea, where she developed an interest in conservation. She maintains an interest in issues related to space technologies and environmental applications.

# Acronyms

| | |
|---|---|
| ABI | Advanced Baseline Imagery |
| ADEOS | Advanced Earth Observing Satellite |
| AFGWC | Air Force Global Weather Central |
| AHMG | Ad Hoc Meteorology Group |
| AIR | All India Radio |
| ALEXI | Atmosphere-Land Exchange Inverse model |
| AMR | Advanced Microwave Radiometer |
| AMSR | Advanced Microwave Scanning Radiometer |
| AMSR-E | Advanced Microwave Scanning radiometer for Earth Observing System |
| APT | Automatic Picture Transmission camera |
| ASCAT | Advanced Scatterometer |
| ATMS | Advanced Technology Microwave Sounder |
| ATN | Advanced TIROS-N |
| ATOVS | Advanced TOVS |
| ATS | Applications Technology Satellite |
| AVCS | Advanced Vidicon Camera System |
| AVHRR | Advanced Very High Resolution Radiometer |
| BC | Before Christ |
| BCC | Beijing Climate Center |
| CASC | China Aerospace Science and Technology Corporation |
| CCD | Charge-coupled device |
| CDR | Climate Data Record |
| CEOS | Committee on Earth Observation Satellites |
| CERES | Clouds and the Earth's Radiant Energy Sensor/System |
| CGMS | Coordination Group for Meteorological Satellites |
| CMA | China Meteorological Administration |
| CMAP | Climate Prediction Center Merged Analysis of Precipitation |
| CMORPH | CPC MORPHing technique |
| CNES | Centre national d'études spatiales |
| COSPAS-SARSAT | Cospas Search and Rescue System |
| CPC | Climate Prediction Center |

| CrIS | Cross-track Infrared Sounder |
| CZCS | Coastal Zone Color Scanner |
| DBS | Direct Broadcast Satellite |
| DCPC | Data Collection and Production Centers |
| DCPs | Data Collection Platforms |
| DCS | Data Collection System |
| DMSP | Defense Meteorological Satellite Program |
| DoC | Department of Commerce |
| DoD | Department of Defense |
| DORIS | Doppler Orbit and Radio Positioning Integration by Satellite |
| DOS | Indian Department of Space |
| DOT | Indian Department of Telecommunications |
| DSAP | Defense Satellite Applications Program |
| DVB | Digital Video Broadcast |
| DWSS | Defense Weather Satellite System |
| ECMWF | European Centre for Medium-Range Weather Forecasts |
| ECVs | Essential Climate Variables |
| ELDO | European Launcher Development Organization |
| ENSO | El Niño/La Niña – Southern Oscillation |
| ENVISAT | Environment Satellite |
| EPS | EUMETSAT Polar System |
| EPS-SG | EUMETSAT Polar System Second Generation |
| ESA | European Space Agency |
| ESMR | Electrically Scanning Microwave Radiometer |
| ESOC | European Space Operations Center |
| ESRO | European Space Research Organization |
| ESSA | Environmental Satellite Service Administration |
| EUMETSAT | European Organization for the Exploitation of Meteorological Satellites |
| EXIS | Extreme Ultraviolet Sensor/X-Ray Sensor Irradiance Sensors |
| FAGS | Federation of Astronomical and Geophysical Data Analysis Services |
| FCI | Flexible Combined Imager |
| FGGE | First GARP Global Experiment |
| FPR | Flat Plate Radiometer |
| FY | Feng-Yun |
| G8 | Group of Eight |
| GAME | GEWEX Asian Monsoon Experiment |
| GARP | Global Atmospheric Research Programme |
| GASP | GOES Aerosol/Smoke Product |
| GAW | Global Atmosphere Watch |
| GCMP | GCOS Climate Monitoring Principles |
| GCOS | Global Climate Observing System |
| GDPFS | Global Data-processing and Forecasting System |
| GEO | Group on Earth Observations |

| GEO | Geosynchronous orbit |
| GEOSS | Global Earth Observation System of Systems |
| GERB | Geostationary Earth Radiation Budget |
| GEWEX | Global Energy and Water Cycle Experiment |
| GFCS | Global Framework for Climate Services |
| GISC | Global Information System Centers |
| GLAS | Geoscience Laser Altimeter System |
| GLM | Geostationary Lightning Mapper |
| GMS | Geostationary Meteorological Satellite |
| GOES | Geostationary Operational Environmental Satellites |
| GOME-2 | Global Ozone Monitoring Experiment-2 |
| GOMS | Geostationary Operational Meteorological Satellite |
| GOOS | Global Ocean Observing System |
| GOS | Global Observing System |
| GPCP | Global Precipitation Climatology Project |
| GPM | Global Precipitation Measurement |
| GPS | Global Positioning System |
| GPSP | Global Positioning System Payload |
| GRAS | Global Navigation Satellite System Receiver |
| GSFC | Goddard Space Flight Center |
| GSICS | Global Space-based Inter-Calibration System |
| GTOS | Global Terrestrial observing System |
| GTS | Global Telecommunication System |
| HIRS | High Resolution Infrared Radiation Sounder |
| HRD | High-rate data broadcasts |
| IASI | Infrared Atmospheric Sounding Interferometer |
| ICESat | Ice, Cloud, and land Elevation Satellite |
| ICSU | International Council for Science |
| IGBP | International Geosphere and Biosphere Program |
| IGOS | Integrated Global Observing Strategy |
| IGY | International Geophysical Year |
| IJPS | Initial Joint Polar-orbiting operational Satellite |
| IMD | India Meteorological Department |
| IMO | International Meteorological Organization |
| INDOEX | Indian Ocean Experiment |
| INSAT | Indian National Satellite |
| IOC | Intergovernmental Oceanographic Commission |
| IPCC | Intergovernmental Panel on Climate Change |
| IPO | Integrated Program Office |
| IR | Infrared |
| IRS | Infra-red Sounder |
| ISCCP | International Satellite Cloud Climatology Project |
| ISES | International Space Environment Service |
| ISRO | Indian Space Research Organization |
| ITCZ | Intertropical convergence Zone |

| | |
|---|---|
| ITOS | Improved TIROS Operational System |
| JANUS | Joint Astrophysics Nascent Universe Satellite |
| JAXA | Japan Aerospace Exploration Agency |
| JCAB | Japan Civil Aviation Bureau |
| JMA | Japan Meteorological Agency |
| JPSS | Joint Polar Satellite System |
| Landsat MSS | Landsat Multispectral Scanner |
| Landsat TM | Landsat Thematic Mapper |
| LEO | Low-Earth-Orbiting |
| LI | Lightning Imager |
| LiDAR | Light Detection And Ranging |
| LIS | Lightning Imager Sensor |
| LRD | Low rate data broadcasts |
| LRPT | Low rate picture transmission |
| LRR | Laser Retroreflector |
| MADRAS | Microwave Analysis and Detection of Rain and Atmospheric Structures |
| MAHASRI | Monsoon Asian Hydro-Atmosphere Scientific Research and Prediction Initiative |
| Megha-Tropiques | Meteorological LEO Observations in the Intertropical Zone |
| MetOp | Meteorological Operational Satellite Program of Europe |
| MEXT | Ministry of Education, Culture, Sports, Science and Technology |
| MFG | Meteosat First Generation |
| MLIT | Japanese Ministry of Land, Infrastructure and Transport |
| MOA | Memorandum of Agreement |
| MOP | Meteosat Operational Program |
| MOS | Multispectral Opto-electronic Scanner |
| MSG | Meteosat Second Generation |
| MSL | Meteorological Satellite Laboratory |
| MSMR | Multi-channel Scanning Microwave Radiometer |
| MSU | Microwave Sounding Unit |
| MSU-M | Multi-spectral Opto-Mechanical Scanner |
| MTP | Meteosat Transition Program |
| MTSAT | Multifunctional Transport Satellite |
| MVISR | Multichannel Visible Infrared Scanning Radiometer |
| NACA | National Advisory Committee for Aeronautics |
| NASA | National Aeronautics and Space Administration |
| NC | National Centers |
| NDBC | National Data Buoy Center |
| NESC | National Environmental Satellite Center |
| NESDIS | National Environmental Satellite, Data, and Information Service |
| NFMOC | Navy Fleet Numerical Oceanography Center |
| NIR | Near Infrared |
| nm | Nanometer |

| | |
|---|---|
| NMS | National Meteorological Services |
| NOAA | National Oceanic and Atmospheric Administration |
| NPOESS | National Polar-orbiting Operational Environmental Satellite System |
| NPP | NPOESS Preparatory Project |
| NSSFC | National Severe Storm Forecast Center |
| NWP | Numerical Weather Prediction |
| NWS | National Weather Service |
| OCM | Ocean Color Monitor |
| OLS | Operational Linescan System |
| OMI | Ozone Monitoring Instrument |
| OMPS | Ozone Mapping and Profiler Suite |
| OSCAT | Ocean SCATterometer |
| OST | Ocean Surface Topography |
| OSTM | ocean Surface Topography Mission |
| OTD | Overshooting Top Detection |
| PIOMAS | Pan-Arctic Ice-Ocean Modeling and Assimilation System |
| POES | Polar-orbiting Operational Environmental Satellites |
| POOMSCOB | Polar-Orbiting Operational Meteorological Satellite COordinating Board |
| PR | Precipitation Radar |
| PSLV | Polar Satellite Launch Vehicle |
| QuikSCAT | Quick Scatterometer |
| R&D | Research and Development |
| RMC | Radiation Measurement Complex |
| ROSA | Radio Occultation Sounder for Atmosphere |
| Roscosmos | Russian Space Agency |
| ROSHYDROMET | Russian Federal Service for Hydrometeorology and Environmental Monitoring |
| RWC | Regional Warning Center |
| SAGE | Stratospheric Aerosol and Gas Experiment |
| SAPHIR | Sondeur Atmospherique du Profil d'Humidite Intertropicale par Radiometrie |
| SAR | Search and rescue |
| SAR | Synthetic Aperture Radar |
| SBUV | Solar Backscatter Ultraviolet Radiometer |
| SCAMS | Scanning Microwave Spectrometer |
| ScaRaB | Scanner for Radiation Budget |
| SCAT | Scatterometer |
| SCOPE-CM | Sustained Coordinated Processing of Environmental Satellite Data for Climate Monitoring |
| SEM | Space Environment Monitor |
| SEM-N | Special Sensor Precipitating Electron and Ion Spectrometer |
| SEVIRI | Spinning Enhanced Visible and InfraRed Imager |
| SIRAL | SAR/Interferometric Radar Altimeter |

| SIT | Strategic Implementation Team |
| SLV | Satellite Launch Vehicle |
| SMC | Space and Missile Systems Center |
| SMMR | Scanning Multi-channel Microwave Radiometer |
| SMS | Synchronous Meteorological Satellite |
| SNPP | Suomi National Polar-orbiting Partnership |
| SPM | Solor Proton Monitor |
| SR | Scanning radiometer |
| SSMIS | Special Sensor Microwave Image/Sounder |
| SST | Sea surface temperature |
| SSU | Stratospheric Sounding Unit |
| SUVI | Solar Ultraviolet Imager |
| S-VISSR | Stretched – Visible and Infrared Spin-Scan Radiometer |
| SXI | Solar X-ray Imager |
| SYNOP | Surface synoptic observations |
| TIR | Thermal infrared |
| TIROS | Television Infrared Observation Satellite |
| TIROS-N | Television Infrared Operational Satellite – Next-generation |
| TOMS | Total Ozone Mapping Spectrometer |
| TOS | TIROS Operational System |
| TOVS | TIROS Operational Vertical Sounder |
| TRMM | Tropical Rainfall Measurement Mission |
| UNEP | United Nations Environment Program |
| UNESCO | United Nations Educational Scientific and Cultural Organization |
| UNFCCC | United Nations Framework Convention on Climate Change |
| USAF | U. S. Air Force |
| USSR | Union of Soviet Socialist Republics |
| USWB | U. S. Weather Bureau |
| UVN | Ultraviolet Visible Near-infrared sounder |
| VAS | Vertical Atmospheric Sounder |
| VHRR | Very High Resolution Radiometer |
| VIRRS | Visible/Infrared Imager/Radiometer Suite |
| VIRS | Visible-Infrared Scanner |
| VLab | Virtual Training Laboratory |
| VNIIEM | All-Russian Scientific and Research Institute of Electromechanics |
| VTPR | Vertical Temperature Profile Radiometer |
| WCRP | World Climate Research Program |
| WCSP | World Climate Services Program |
| WEFAX | Weather facsimile data |
| WIGOS | WMO Integrated Global Observing System |
| WIS | WMO Information System |
| WMO | World Meteorological Organization |
| WWW | World Weather Watch |

# Chapter 1
# Introduction to Meteorological Satellites

*Some things are only capable of being done in space. Examples
of that are looking at our Earth from that far away, and
understanding the entire processes of storms and weather
patterns, and oceans, and coastlines.*
—Laurel Clark (STS-107 Preflight Crew Interview, 2002)

Meteorological services are now utilized by every nation in the world. These services include weather forecasts, public warnings, and providing products and information for the purposes of protection and safety. Weather forecasting and severe weather warnings are essential to the success of the public and private sectors, including business and commerce, agriculture, forestry, marine and fisheries, the airline industry, military applications, and urban infrastructure management. The timely collection of data related to current weather or information required for forecasting future conditions is essential to any weather forecast. Indeed, a meteorologist's forecast will only be as good as the data and knowledge that is available.

World weather observations are carried out by various approaches, including aircraft, ships, and ocean buoys with observing capability, weather balloons, and upper atmospheric and ionospheric sounding capabilities. Weather forecasts are frequently based on information gathered by weather radar, Doppler wind profilers, laser cloud ceilometers, and ionospheric sounding systems. The bulk of weather data in all environmental domains is currently remotely sensed with most observations and surveillance achieved by satellite-based instrumentation (Fig. 1.1).

The routine use of meteorological satellite data is now entirely accepted. Yet, it was totally new a half a century ago. Before the space age that began with the launch of Sputnik-1, on October 4, 1957, many people were quite skeptical that meteorological satellites could become a significant aid to weather forecasting. In fact, the launch of Sputnik spurred new military, political, scientific, and technological developments. As weather observing tools, satellites offered new capabilities for meteorologists, such as the ability to view Earth as a whole in a short period of time. Satellites could also provide very useful day-to-day monitoring capabilities, and information about remote regions where traditional data sources were not previously available. Collectively, weather satellites from the United States, Europe, India, China, Russia, and Japan can provide nearly continuous observations for a global weather watch.

**Fig. 1.1** Image of Earth created using data from four different satellites. Satellite data and imagery provide a comprehensive view of Earth's systems and climate. (Courtesy of NASA) [1]

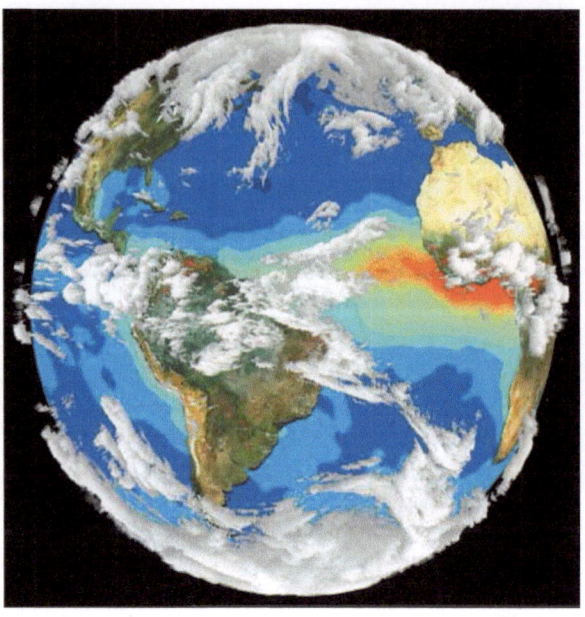

The first weather satellite was successfully launched by NASA on April 1, 1960. This satellite was called the Television Infrared Observation Satellite (TIROS-1). It released fuzzy images of thick bands and clusters of clouds over the United States (Fig. 1.2). This satellite would forever change weather forecasting and climate research in the field of Earth system science. The United States launched ten TIROS satellites in total. TIROS satellite data was directly transmitted in real-time to forecasting centers and ground stations within signal range of the satellite. By 1965, TIROS imagery had been combined to generate the first global view of worldwide weather. The success of these meteorological observations proved to be effective for meteorological and environmental surveillance, paving the way for the Nimbus program, which is the heritage of most Earth-observing satellites. These Nimbus satellites were launched by the U.S. National Aeronautics and Space Administration (NASA) and the National Oceanic and Atmospheric Administration (NOAA) [2].

Weather satellites are categorized into two types, polar-orbiting and geostationary satellites, which collectively provide a complete global weather monitoring system. Polar-orbiting weather satellites travel in a north to south (or vice versa) low-orbit path that passes over both poles to collect ocean, land, and atmospheric data across the entire globe, essential for longer-term forecasting. Geostationary satellites, orbiting together at the same rate as Earth's rotation, provide constant observation for short-range warning and so-called "now-casting." These satellites provide key data related to severe weather events, such as hurricanes, heavy rainfall, and tropical storms. Orbital characteristics for observing Earth help to determine the types of synthetic applications for which a weather satellite is best suited.

**Fig. 1.2** One of the first images captured by the TIROS-1 satellite on April 1, 1960. (Courtesy of NASA) [3]

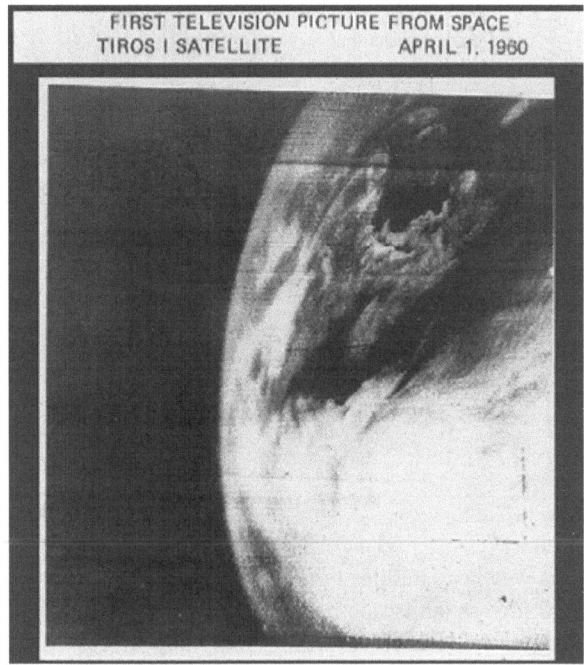

FIRST TELEVISION PICTURE FROM SPACE
TIROS I SATELLITE          APRIL 1, 1960

End uses for weather forecasts are wide-ranging. Remote meteorological observations have been applied extensively for oceans and marine applications, such as ship safety, fishing, weather forecasting, and global-scale studies of climate and sea conditions. Satellite observations of ocean conditions provide benefits to marine fisheries, especially for monitoring seawater temperature fluctuations, which significantly affect fish species at different stages of their life cycles. Monitoring ocean surface temperature variations with satellite observations has become important for studying the relationship between the environment and fish distribution, aggregation, migration, and schooling behavior. Another parameter used to assess fishing resources is phytoplankton biomass, which is the primary source of aquatic food and the primary agent for production in the oceans. For example, near-surface phytoplankton pigment concentration was early on estimated using imagery from the Nimbus-7 Coastal Zone Color Scanner (CZCS).

Weather satellites are also used to collect longer term data related to climate change and to assist with environmental protection. Such environmental applications are far-ranging, since on-board sensors can detect a variety of phenomena, including the location and movement of cloud systems, particles from fires, snow and ice cover, sand and dust storms, ocean currents, and pollution. Since airborne particles with a diameter of less than 10 μm can penetrate deep into human lungs and potentially cause serious health problems, the current Geostationary Operational Environmental Satellite (GOES) series is being used to forecast national air quality, by providing an aerosol product known as the GOES Aerosol/Smoke

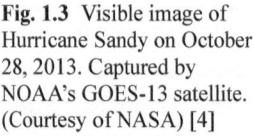

**Fig. 1.3** Visible image of Hurricane Sandy on October 28, 2013. Captured by NOAA's GOES-13 satellite. (Courtesy of NASA) [4]

Product (GASP). The GASP provides a good proxy for national pollution monitoring. Weather satellites have also been used to track other environmental phenomena, such as landscape fire detection. Environmental data that is collected over the longer term can be applied to the study of global warming and climate change. Other types of scientific satellites can be used to study so-called "space weather" and phenomena such as solar flares, coronal mass ejections, and cosmic radiation, but these types of applications are only briefly discussed in this book. Another book in this series entitled *Orbital Debris and Other Hazards from Outer Space* address some of these issues in greater detail.

Timely and reliable meteorological data not only improves the quality of life but also contributes to saving lives. National Oceanic and Atmospheric Administration (NOAA) satellites can pinpoint downed pilots, shipwrecked mariners, or stranded hikers by detecting a distress signal from an emergency beacon activated onboard an aircraft/boat or a handheld personal locator, relaying the information to first responders on the ground. NOAA's polar-orbiting and geostationary satellites, along with Russia's COSPAS spacecraft, form part of the International COSPAS-SAR-SAT program, a search and rescue satellite-aided tracking system for distress alert detection and information distribution. This system consists of both ground and space segments, including a network of satellites, ground stations, mission control centers, and rescue coordination centers. Accurate information collected on wind patterns can aid the National Hurricane Center when producing a more reliable forecast of a hurricane's track, which may aid disaster management and evacuation planning of thousands of families (Fig. 1.3).

Integration of satellite imagery and weather data has become increasingly important for improving agricultural management decisions. A thorough knowledge of the moisture and temperature conditions of a field improves agricultural forecasting and now-casting, which provides powerful agronomic modeling and assessment tools that can be developed for field and sub-field management. For example, weather satellites can help forecast low temperatures in the fall and winter, which could aid farmers in making correct decisions regarding deployment or non-deployment of freeze prevention methods for crop protection. Early warning for drought preparedness and management is critical to the agricultural industry, as well as long-term monitoring of global sea-surface temperatures that signal El Niño and La Niña events that could potentially impact vegetation.

The analysis of satellite imagery for identifying crop variability has been exploited since the availability of Landsat MSS/TM data in the mid 1970s, while operational weather satellite data have been used for agricultural applications since the late 1980s. Satellite imagery is valuable for drought warning, since radiance measurements by space sensors, especially by the Advanced Very High Resolution Radiometer (AVHRR) on NOAA polar-orbiting satellites, respond closely to alterations in leaf chlorophyll, thermal conditions, and moisture content. Weather satellite data combined with historic, current, and forecasted weather observations can be integrated as a crop management tool, monitoring the effectiveness or need for irrigation, treatments for diseases and pests, and the addition of nutrients or fertilizers.

In addition to the societal benefits of services and data provided by meteorological satellites, there are also immense economic benefits of the weather forecasts and information derived from satellite observations. This is especially apparent when considering the total value of annual weather-related losses and damages occurring on a worldwide scale. According to NOAA Chief Economist Rodney Weiher, the economic benefits of providing reliable warnings of geomagnetic storms to the electric power industry in the United States would be approximately $ 450 million over three years alone without considering the economic impacts of a widespread blackout [5].

National welfare is increased by allocating investment, labor, capital, and other resources to weather satellite operations. For example, a relatively accurate 24-hour forecast of heavy rain and violent storms along a cold front enables airline dispatchers to reroute their aircrafts in advance to avoid costly flight delays. In the electrical energy industry, through direct investment in meteorological data and information, monitoring and forecasting space weather can assist electric grid operators on Earth to take steps to avert failures. Space weather in the form of magnetic disturbances caused by coronal mass ejections from the Sun creates geomagnetically induced currents that disturb the electric power grid, which can cause significant economic impacts on electrical energy distribution. This information can assist in planning the timing of satellite launches, warning astronauts onboard the International Space Station (ISS) to take shelter, and preventing damage to electronics and infrastructure.

Meteorological satellites help save lives during disasters. They are key to providing severe weather alerts and advisories issued by national weather services when hazardous weather is expected. For example, near-real time satellite imagery was used by the National Severe Storms Forecast Center in Kansas City to issue a tornado warning in May 31, 1985, extending from eastern Ohio into New York and Pennsylvania. Satellite observations of developing storms allowed early actions and initiated warning systems before the category F5 tornado struck the town of Newton

Falls, Ohio, avoiding fatalities in the area. Since that time, this type of predictive action based on satellite data has been repeated many times over in locations all over the world.

The next generation geostationary weather satellite series, called the GOES-R, is to be launched by 2015. This satellite that will monitor lightning strike patterns will be able to provide even earlier and more accurate warnings for severe weather. This will lead to improved severe weather detection. For the first time, GOES-R will enable scientists to detect lightning occurring inside storm clouds, thus enabling more accurate tracking of how developing storms move and intensify before and during severe weather. Its instruments will also allow researchers to observe cloud and surface changes in greater detail and to monitor solar radiation for improved forecasting of space weather.

The surveillance of Earth and its environment by meteorological satellites form the backbone of today's weather forecasts, climate research, and environmental assessments that result in public awareness and preparedness. Satellites launched by the United States, Europe, Russia, India, China, and other countries increase the potential for greater accuracy and additional capability for life-saving weather forecasts, ocean temperature measurement, hazards planning and severe weather detection. Satellites offer increasingly sophisticated imaging systems for day and night global weather-monitoring capabilities, sending back an endless stream of information about ocean, land, atmosphere, and space conditions.

Worldwide weather data collection networks allow effective sharing of meteorological data on a global scale (e.g., World Meteorological Organization, European Centre for Medium Range Weather Forecasting), facilitating free and unrestricted exchange of data, products, and services among member nations. Moreover, the Coordination Group for Meteorological Satellites (CGMS) was established in 1972, supporting the exchange of technical information on geostationary and polar orbiting meteorological satellite systems. However, despite coordinated efforts for data sharing, there are still large areas of the world for which little day-to-day weather information is available. This includes remote locations, such as the oceans, polar regions, deserts, and forested areas.

The meteorological satellite observation data currently available also originate from a wide variety of federal and non-federal sources, including government agencies, private companies, research organizations, and universities. There is a need to link meteorological observations collected by satellites and traditional methods together into integrated databases that are easy to use and have a standard format that provides accurate and timely information. Integrated datasets are required to improve forecasts of severe weather and natural hazards, enhancing forecasting models and creating a more complete picture of our global climate system.

The purpose of this book is to provide a brief but comprehensive guide to weather satellites and their Earth applications. This publication benefits the readership by providing a ready and quick reference for comprehensive information about meteorological satellites. It is intended to serve as a useful handbook for a broad audience, including students, academia, private consultants, engineers, scientists, and teachers. It is a multi-disciplinary reference that spans the fields of engineering, meteorology and climatology, Earth and atmospheric sciences, Earth observation, remote sensing, and physical geography.

Here, now, is a review of the organization of the book.

This first chapter of *Meteorological Satellite Systems* reviews basic concepts behind space meteorology technologies and applications. Chapter 2 reviews historical attempts to monitor and understand weather patterns and climates. It notes the first efforts to use balloons and aircraft and then the evolution of geostationary and polar-orbiting satellites to monitor weather patterns over land and the oceans.

Chapter 3 explains the various types of meteorological satellites and their functions and applications. It describes how different types of satellites are used on an integrated basis to develop both short-term and longer-term weather forecasts, as well as the evaluation of sensing technologies to allow for high resolution data collection and merging of data from satellites of different countries.

Chapter 4 details the evolution of U.S. meteorological satellites and the complementary roles of NASA, NOAA, and the U.S. Department of Defense in developing technologies for the analysis and use of data from meteorological satellites.

Chapter 5 provides an overview of the European meteorological satellite system, which has a very extensive meteorological satellite capability in both geostationary and polar-orbiting satellites, sometimes on a cooperative basis with the United States. The nature of the satellites and cooperative programs are discussed.

Chapter 6 introduces the meteorological satellite systems of other countries, including Russia, China, Japan, and India, while comparing their strengths and capabilities. Together, these satellites provide important input into the World Weather Watch and global weather forecasting.

Chapter 7 discusses international collaboration in meteorological satellite systems, including the role of the U.N.'s World Meteorological Organization, the World Weather Watch, and other key collaborative efforts for weather data sharing. These international coordination efforts include the Group on Earth Observations (GEO) and the Committee on Earth Observation Satellites (CEOS). These activities have a key goal of guaranteeing the continuous long-term availability to all nations of all space-observable Essential Climate Variables (ECVs), as defined by the Global Climate Observing System (GCOS). It also touches on the U. N. Environmental Program in the context of how meteorological satellite systems also help with environmental monitoring and charting climate change.

Chapter 8 examines future capabilities of meteorological satellites that are evolving to allow for accurate monitoring of violent storms, such as lightning strike monitoring that indicate storm vectors, and the most up-to-date means of integrating data from GEO satellites and polar-orbiting satellites.

Chapter 9 explains how meteorological satellites are used to monitor ozone holes, weather and rain patterns related to El Nino and La Nina events, and other longer term changes in weather patterns related to climate change.

Finally, Chap. 10 concludes with a "Top Ten List" summarizing key facts to know about meteorological satellites.

# References

1. NASA: http://history.nasa.gov/SP-4312/ch5.htm Wallace, L.E. Dreams, Hopes, Realities: NASA's Goddard Space Flight Center, the first forty years. (SP–4312). Accessed 12 Aug 2013
2. NASA—Nimbus: 40th Anniversary http://earthobservatory.nasa.gov/Features/Nimbus. Accessed 12 Aug 2013
3. NASA: http://www.noaanews.noaa.gov/stories2010/20100401_tiros.html. Accessed 12 Aug 2013
4. NASA: http://www.nasa.gov/mission_pages/hurricanes/archives/2012/h2012_Sandy.html. Accessed 12 Aug 2013
5. Teisberg, T.J., Weiher, R.F.: Valuation of geomagnetic storm forecasts: an estimate of the net economic benefits of a satellite warning system. J. Policy. Anal. Manag. **19**(2), 329–334 (2000)

# Chapter 2
# History and Background

> *Man must rise above the Earth—to the top of the atmosphere and beyond—for only thus will he fully understand the world in which he lives.*
>
> —Socrates

## A History of Weather Forecasting

It is human nature to want to find out about our surroundings, to explore our neighborhood, our planet Earth, and beyond. Until the twentieth century, viewing Earth from a space-based perspective could only be accomplished by imagination. From ancient times, astronomers have looked up at the sky, recorded their observations, and made up stories about how the universe was created and what it was like. Ancient Greeks were more aware of the truth of their surroundings than other cultures in that time period. They helped to discover that Earth was a sphere and developed observational and mathematical techniques to measure the circumference of the planet. With increasingly powerful ground-based telescopes came the discovery of the Milky Way and other galaxies and our understanding that the universe is expanding.

Space exploration has improved our understanding of Earth as a celestial object in its own right. The urge to view and explore Earth from above is perhaps also intrinsic to human nature—for example, when a mountaineer wonders how Earth appears to a hawk soaring above him, or when the first explorers tried to reach the ends of the Earth. The practice of Earth observation involves the gathering of information about the planet's physical, chemical, and biological systems, usually by remote sensing systems, which have grown technologically more and more sophisticated over time. The famous "Big Blue Marble" photograph of Earth, taken in 1972 by astronauts onboard Apollo 17, demonstrated the dramatic impact of viewing Earth from space (Fig. 2.1). This emphasized the importance of minimizing the negative impact of modern human civilization to improve social and economic well-being.

The art of weather forecasting, which began with early civilizations and was based on recurring astronomical and meteorological events, was used to monitor and predict seasonal changes in the weather. In 650 BC, the Babylonians attempted to predict short-term weather based on cloud patterns and astrological observations,

S.-Y. Tan, *Meteorological Satellite Systems,* SpringerBriefs in Space Development, DOI 10.1007/978-1-4614-9420-1_2, © The author 2014

while Chinese weather prediction dates back to 300 BC, when annual calendars were developed according to repeated patterns of weather events. This experience accumulated over generations to produce weather lore. In about 340 BC, Aristotle described weather patterns in a treatise entitled *Meteorologica*. This writing contained Earth science theories, such as on cloud formations, wind, rain, and other weather phenomena. This led to his pupil, Theophrastus, compiling *The Book of Signs*, which documented weather lore and forecast signs. These texts served as definitive weather forecasting references for more than 2,000 years and helped to establish meteorology as a distinct discipline of study. Weather forecasting advanced little from these ancient times until the Renaissance, despite many of Aristotle's claims being erroneous.

## Early Meteorological Instrumentation

Over the centuries, it became apparent that forecasts based on weather lore, philosophical speculations, and personal observations alone were not always reliable [2]. In order to advance knowledge and understanding of the atmosphere, instruments were needed to measure properties, such as moisture, temperature, and pressure. The first device to measure the humidity of air, called the hygrometer, was invented by Leonardo da Vinci in the fifteenth century. About 1593, Galileo Galilei, often deemed the father of modern observational astronomy, invented an early thermometer for temperature measurement using the expansion and contraction of air in a bulb to move water in an attached tube. His student Evangelista Torricelli invented

**Fig. 2.2** A 1783 drawing of the first hot-air balloon, invented by the French brothers Étienne and Joseph Montgolfier. (Courtesy of *The New York Times*) [4]

the barometer for measuring atmospheric pressure in 1643. In subsequent centuries, these meteorological instruments were refined and improved, and were being applied in association with observational platforms, such as balloons and aircrafts, for taking atmospheric meteorological measurements.

The modern age of weather forecasting began with the invention of the electric telegraph in 1835, which allowed for routine and almost instantaneous transmission of weather observations. It was possible to develop crude weather maps and to study surface wind patterns and storm systems. Synoptic weather forecasting was made possible by the compilation and analysis of data collected simultaneously from weather observing stations and conveyed across the globe via telegraph in the 1860s [3]. Data collected by land locations are now conveyed worldwide via phone lines or wireless technology, enabling information to be communicated quickly for weather forecasts and studies of the atmosphere and climate.

With advancements in meteorological instrumentation in the eighteenth century came experimentation of different airborne platforms to measure physical properties of the atmospheric column, including pressure, temperature, wind speed, wind direction, and other properties. A significant historical development was the invention of the first balloon in 1783 by Étienne and Joseph Montgolfier (Fig. 2.2). They experimented with hydrogen-filled paper bags. Their experiments led to the correct notion that a buoyant force should cause ascent of the bags, if the inside gas was lighter than air. However, since gas diffused out quickly and hydrogen was produced in small quantities, they subsequently tried 'gas' produced by the combustion of a mixture of moistened straw and wool. This produced the first hot-air balloon in the world, which attained a height of about 1,950 m.

In the same year, J.A.C. Charles and the Robert brothers designed and constructed a hydrogen-filled balloon, but inflation was achieved only with great difficulty over a period of four days. Balloon flights began carrying animals and then subsequently men. Furthermore, balloons were improved to descend and ascend at will, and in time were improved with safe landing devices and better direction control. As techniques of the manned balloon evolved rapidly, it offered the possibility to investigate

Earth's atmosphere. The first manned hydrogen balloon flight conducted by J.A.C. Charles and the Robert brothers carried a barometer and thermometer to measure the pressure and the temperature of the air, making it the first balloon flight to provide atmospheric meteorological measurements above Earth's surface [5].

In 1784, the first balloon flight for the scientific purpose of studying environmental conditions was conducted by the American physician, J. Jeffries, along with J.P. Blanchard. Experiments were carried out along a flight path from London to Dartford [6]. Also in 1784, Charles rose again and measured temperature variation along with altitude in the atmosphere. After 1850, the application of balloons for measuring meteorological parameters was widely practiced. To monitor favorable weather conditions for balloon ascent, Charles also inaugurated the practice of launching a small pilot balloon prior to flight in order to determine the wind vector at different altitudes.

Thereafter, pilot balloons were superseded by free-flight sounding balloons, which carried sensors and telemetry transmitters. Sensors launched by weather balloons to measure atmospheric profiles of pressure, temperature, and relative humidity are carried in a unit commonly referred to as a radiosonde. Usually contained in a small, expendable instrument package suspended below a large balloon, the radiosonde provided an efficient way to systematically and regularly measure various atmospheric parameters to heights of over 100,000 feet, without the necessity of considering weather conditions. Nowadays, received meteorological information is transmitted to a ground-based station for data users via a radio transmitter and radiosondes are still used by national weather services for capturing high vertical resolution flight data.

Kites were frequently used for capturing meteorological observations in the second half of the nineteenth century. A meteographic device, which is a chart recorder for measuring humidity and temperature, was usually attached. However, kites were linked to the ground and highly unstable due in windy conditions. During World War I, meteorological observations from kite flights were largely substituted for by aircraft. Flying weather forecasts for aircraft were not required prior to World War I, since pilots mostly flew at low altitudes.

During wartime, however, aircraft were often required to fly in clouds, in bad weather conditions, and at high altitudes. Meteographs were often mounted on the wings of military aircraft to obtain meteorological information for monitoring flying conditions. Observations were recorded on a cylindrical chart that was retrieved after the landing of the aircraft, and meteorological parameters were read from the chart. Pilots were often required to reach a flying altitude of at least 13,500 feet, where they could black out from lack of oxygen, making this a very dangerous enterprise [7]. It was often impossible to fly in bad weather, which unfortunately was when observations were needed the most.

Subsequent advances in the use of unmanned balloons made it possible to sound the atmosphere. For example, Colonel William Blaire in the U.S. Signal Corps performed primitive experiments with weather measurements from a balloon, while the first really useful radiosonde was invented in France by Robert Bureau in 1929. This device sent precise encoded telemetry from weather sensors to the ground.

**Fig. 2.3** Early launch of radiosondes developed by the U.S. Bureau of Standards in 1936 (*left*). Army Air Force meteorologists preparing a hydrogen-filled balloon equipped with a radiosonde in 1944 (*right*). (Courtesy of NOAA's National Weather Service Collection) [9]

Subsequent developments enabled radiosonde instruments to become smaller, lighter, and more accurate (Fig. 2.3). Radiosondes have also been used for exploring atmospheres of other planets, such as in the Soviet Union's Vega program, where probes dropped radiosondes to study the atmosphere of Venus. Up to present, radiosondes are still launched worldwide year-round. The National Weather Service in the United States releases about 75,000 radiosondes each year, not including soundings made by military facilities and for other specialized scientific purposes. Collective agreements have formed a global radiosonde station network worldwide (about 900 stations) that make an average of 1,209 soundings each day to support weather forecast activities [8].

## The Evolution of Weather Satellites

The vast arrays of radiosonde stations, weather reconnaissance aircraft, and other newly developed observing systems have provided a vast amount of information about meteorological parameters and weather conditions. Sensors measuring atmospheric constituents directly, such as thermometers, barometers, and humidity sensors, have been sent aloft on balloons, rockets, or dropsondes for many years. Although precise in their measurements, these instruments have limited capabilities to provide regional or global coverage, which is necessary for making accurate weather forecasts. The global network of radiosonde observing stations tend to have a highly concentrated dispersion in the northern hemisphere temperate zone land masses of North America, Europe, and Asia, whereas the density of observations for the southern hemisphere, tropical regions, the Arctic, and most of the northern

Pacific is relatively sparse. Consequently, there is a high degree of uncertainty with tracking storms over the north Pacific Ocean.

Since the Earth's atmosphere is a single and closely interacting mass of air, disturbances can propagate throughout at a speed much faster than winds. Hence, real-time and synoptic monitoring of large areas of the Earth is necessary for improved meteorological data collection. Extended and long-range weather forecasts require data to be collected and distributed globally. Earth-observing satellites are able to collect meteorological data at synoptic scales and in remote locations, tracking cloud cover, relative motion of storm systems and the jet stream, and maximum heights of clouds and vertical temperature profiles. Satellite imagery can identify cloud patterns associated with different types of weather conditions and patterns (e.g., spiral cloud patterns and convective cells), which are difficult to capture and monitor using conventional weather observations alone. Developments in satellite technologies have resulted in enormous improvements in the accuracy of weather forecasting. Satellites have particularly provided routine access to observations and data from remote areas of the globe.

## Polar-Orbiting Weather Satellites

Before the first meteorological satellite was launched, rockets carrying cameras were used to determine the attitude of the nose cone in space with other instruments used to photograph Earth below and to observe cloud formations. However, short-lived rocket observations proved to be insufficient for meeting meteorologists' requirements for weather information. A transition to long-duration orbiting satellite observation platforms was required. On April 1, 1960, NASA's TIROS-1 was the first successfully launched photographic weather satellite with an inclination of 48° (Fig. 2.4). As an experimental spacecraft, TIROS-1 operated for only 78 days, but it sent back nearly 23,000 pictures of Earth and its ever-shifting cloud cover from an altitude of about 700 km.

Sightings from the surface had not prepared meteorologists for the interpretation of the cloud patterns [10]. TIROS images were able to provide the visible expression of invisible air masses, frontal systems, and wind fields. After the launch, military forecasters initiated operational use of panoramic cloud images from weather satellites [11]. TIROS-9 and TIROS-10, launched in 1965, were the first two polar-orbiting meteorological satellites in the TIROS program. Previous satellites had operated at an inclination of 48 or 58°, which were not polar-orbiting.

The TIROS program not only contributed to the development of a meteorological satellite information system but also enabled testing of various design issues for spacecraft. This program thus provided tests of sensing instruments, data collection processes, and of various operational parameters. TIROS-1 through -10 were known as the first generation of American weather satellites and proved to be extremely successful. This first series of meteorological satellites paved the way for further exploration using space-borne weather prediction and monitoring devices.

**Fig. 2.4** The TIROS-1 satellite that was NASA's first experimental weather satellite. (Courtesy of NASA) [12]

After the experimental launches of the TIROS program, the ESSA satellite program was initiated. ESSA-1 launched successfully in 1966, becoming the first dedicated operational meteorological satellite. Its primary mission was to provide cloud-cover photography to the U.S. National Meteorological Center (now called the National Centers for Environmental Prediction) for the purpose of preparing operational weather analyses and forecasts [13]. This resulted in a combined effort from NASA, ESSA, the U.S. Weather Bureau, and the National Meteorological Center. This operational program consisted of nine satellites (ESSA-1 to ESSA-9) that were launched from 1966 to 1969. Similar to the TIROS satellite series, ESSA-1 also had a spin-stabilized design. Advances in technology enabled more information of a much wider scope and better resolution to be gathered. These satellites transmitted thousands of images back to Earth over a period of almost four years.

In parallel with the operational ESSA series, NASA developed and maintained a research series of seven Nimbus satellites from 1964 to 1978. One of the major goals was to serve as a test bed for future operational polar-orbiting instruments and advanced systems for sending and collecting atmospheric science data. Seven satellites were launched over a fourteen-year period and the Nimbus program became the primary research and development platform for Earth observation. This Nimbus research program formed the heritage of most NASA and NOAA satellites [14]. It carried a variety of instruments, including microwave radiometers, atmospheric sounders, ozone mappers, CZCS, and infrared radiometers, thus providing a significant source of atmospheric chemistry, physics, and climatic data. Nimbus missions contributed to significant advancements in knowledge about weather forecasting, Earth's radiation budget, ozone layer, and sea ice.

However, a demand for more accurate weather analysis meant that higher spatial and temporal resolution data was required. NASA's next step in improving the operational capability of weather satellite systems was the NOAA ITOS (Improved TIROS Operational System), and then the TIROS-N/NOAA program (Television Infrared Operational Satellite—Next-generation). This series of satellites provided higher resolution imaging, improved observations, and expanded operational capabilities. This included more day and night quantitative environmental data on local and global scales with technologically superior instrumentation than earlier TIROS satellites [15]. Moreover, the satellites carried AVHRR and the TIROS Operational Vertical Sounder (TOVS), as well as a fully digital system. A more detailed historical overview of the development of the U.S. meteorological satellite program is provided in Chap. 4.

When the more advanced TIROS-N series satellites were launched between 1978 and 1981, the name of the spacecraft constellation was changed to Polar-Orbiting Operational Environmental Satellites (POES) due to their polar-orbiting nature. POES satellites orbit Earth at an altitude of about 800 km and circle the poles once every 102 min, completing roughly 14.1 orbits per day. This type of low Earth and polar orbit is unfortunately quite popular for many applications, and thus, these orbits are becoming increasingly congested by many active and defunct satellites and various types of orbital debris.

Since the number of daily orbits is not an integer, the ground tracks do not repeat on a daily basis, although local solar time of each satellite's passage is essentially unchanged. This system includes both morning and afternoon satellites, providing global coverage four times daily. POES satellites are able to collect global data on a daily basis for a variety of environmental applications, including weather forecasting, climate research, global sea surface temperature measurements, atmospheric soundings of temperature and humidity, ocean dynamics research, volcanic eruption and forest fire monitoring, global vegetation analysis, search and rescue, and many other applications [16]. POES data offer the benefit of continuous and reliable data products that can have many human and environmental applications.

## Geostationary Weather Satellites

On December 7, 1966, the Applications Technology Satellite (ATS-1) was launched—the first of six spacecrafts used to test the feasibility of placing a satellite into geosynchronous orbit (GEO). It was originally intended to be a communications satellite, but also provided a platform for meteorological and navigation equipment. The satellite was designed to test GEO orbit techniques and applications at this special orbit that circles Earth once a day (Fig. 2.5). This orbit is exactly 22,236 miles (or 35,786 km) above Earth's surface. In this GEO orbit within Earth's equatorial plane, a satellite can transmit information to surface ground stations that are constantly pointed toward the satellite without tracking. This makes the orbit excellent for telecommunications, video broadcasting, and Earth observation [17].

**Fig. 2.5** The Synchronous Meteorological Satellite (SMS-1) was the first geosynchronous weather satellite, which later evolved into the GOES program (*left*). The sixth Applications Technology Satellite (ATS-6) undergoing prelaunch testing at Cape Canaveral, Florida (*right*). (Courtesy of NASA) [19]

Temporally continuous geostationary satellite images offer the ability to track clouds. With this information wind speeds at cloud altitude can be inferred [18]. Research into tracking clouds using image sequences began almost immediately, especially with the successful operation of the ATS-1 spin-scan camera. This imaging device provided the first high quality cloud-cover pictures and afforded a continuous watch of global weather patterns. The success of meteorological experiments carried aboard the ATS-1 and ATS-3 satellites led to NASA's development of two weather satellites designed specifically to make atmospheric observations, called the Synchronous Meteorological Satellites (SMS-1 and SMS-2). These geosynchronous meteorological satellites were launched in 1974 and 1975 (Fig. 2.5).

In 1975, the GOES program was formally initiated with the first operational spacecraft GOES launch. Its ability to orbit in sync with Earth's rotation, along with the polar-orbiting satellites and Defense Meteorological Satellite Program (DMSP) polar-orbiting satellites enhanced NOAA's forecasting capabilities. In the following year, 1977, Japan and Europe launched their first GEO weather satellites. These were respectively, the Geostationary Meteorological Satellite (GMS-1) and the European Meteorological Satellite (Meteosat-1). Meteosat-1 provided visible imagery with a spatial resolution of 2.5 km, and infrared window band imagery and water vapor band imagery, both at 5 km spatial resolution. Its water vapor imagery provided a very different view of Earth. It primarily observed upper tropospheric humidity and high cloud features, which indicated synoptic scale circulations. In 1979, three GOES and one METEOSAT satellites were used as part of a Global Atmospheric Research Program (GARP) to define global atmospheric circulations.

Since satellite systems are beyond the resources of most individual countries, the European Organization for the Exploitation of Meteorological Satellites (EU-METSAT) was established in 1986. This intergovernmental organization was governed by a council of 18 member states, 26 member states, and 5 cooperating states as of 2010, with the purpose of establishing, maintaining, and exploiting European systems for operational meteorological satellites and to contribute to the monitoring and understanding of climate change. EUMETSAT operates a system of meteorological satellites for monitoring the atmosphere and ocean and land surfaces, delivering a continuous stream of weather and climate-related satellite data, images, and products on which National Meteorological Services of member states depend. This information is critical for monitoring potentially dangerous severe weather conditions, issuing timely forecasts and warnings, and helping to protect human life and property. It is also critical to industries, including aviation, maritime, road traffic, agriculture, and construction, among others. National meteorological satellite programs are covered in detail in Chap. 4, 5, and 6, while EUMETSAT is discussed in depth in Chap. 5.

# References

1.  NASA: http://visibleearth.nasa.gov/view.php?id=55418. Accessed 12 Aug 2013
2.  Goldstein M.: "The complete idiot's guide to weather". Alpha/Penguin Group Inc., Indianapolis, USA (2002)
3.  NASA: http://earthobservatory.nasa.gov/Features/WxForecasting/wx2.php. Accessed 12 Aug 2013
4.  The New York Times: http://graphics8.nytimes.com/images/2006/10/13/sports/13vecsey.1.600.jpg. (2006). Accessed 12 Aug 2013
5.  Pfotzer, G.: "History of the use of balloons in scientific experiments". Space. Sci. Rev. **13**(2), 199–242 (1972)
6.  Torenbeek, E., Wittenberg, H.: Flight physics: essentials of aeronautical disciplines and technology, with historical notes. Springer, New York (2009)
7.  Hughes, P., Gedzelman, D.: The new meteorology. Weatherwise. **48**, 26–36 (1995)
8.  Dabberdt, W.F., Shellborn, R., Cole, H., Paukkunen A., Horhammer, J., Antikainen, V.: Radiosondes. http://www.eol.ucar.edu/homes/juhong/Ecy-radiosonde/pdf (2002). Accessed 12 Aug 2013
9.  NOAA's National Weather Service (NWS) Collection: http://www.photolib.noaa.gov/nws/kite1.html. Accessed 12 Aug 2013
10. NASA: Missions. http://science.nasa.gov/missions (2012)
11. Hughes, P.: The view from space. Weatherwise. **48**(60), 60–62 (1995)
12. NASA: http://eospso.gsfc.nasa.gov/missions/television-infrared-observation-satellite-program. Accessed 12 Aug 2013
13. NASA, ESSA: http://science1.nasa.gov/missions/essa/. Accessed 12 Aug 2013
14. NASA, Nimbus: http://science1.nasa.gov/missions/nImbus/. Accessed 12 Aug 2013
15. NASA, TIROS: http://science1.nasa.gov/missions/tiros/
16. NOAA—Polar-orbiting operational environmental satellites. http://www.ospo.noaa.gov/Operations/POES/index.html. Accessed 12 Aug 2013
17. NASA: ATS http://science1.nasa.gov/missions/ats/. Accessed 12 Aug 2013
18. Johnson, D.S. Development of the operational program for satellite meteorology. NASA Conference Publication. **2257**, 34–40 (1982)
19. NASA: http://history.nasa.gov/SP-4312/ch5.htm *Wallace, L.E. Dreams, Hopes, Realities: NASA's Goddard Space Flight Center, the first forty years.* (SP–4312). Accessed 12 Aug 2013

# Chapter 3
# Examining the Tools of Space Meteorology

> *On a visit to the space program, President Kennedy asked me*
> *about the satellite. I told him that it would be more important*
> *than sending a man into space. "Why?" he asked. "Because,"*
> *I said, "this satellite will send ideas into space, and ideas last*
> *longer than men.*
>
> —Newton N. Minow

## Types of Meteorological Satellites

In general, meteorological satellites fall into two categories defined by classes of orbit, namely, polar-orbiting and geostationary. These orbits are illustrated in Fig. 3.1. Since both types of satellites produce different types of data and information, they are better suited for different applications. Combining usages and integrating data products from both systems have also been extensively implemented with new data fusion techniques.

Geostationary satellites orbit Earth over the equator at an altitude of approximately 36,000 km (22,370 miles) above Earth (Fig. 3.1a). At this particular altitude, satellites complete one orbit in 24 h synchronized with Earth's rotation about its own axis and its movement around the Sun. Consequently, they remain over the same location on the equator, thereby enabling continuous surveillance of Earth and space. The primary advantage of geostationary satellites is the high temporal resolution of their data.

Due to their high altitude, some geostationary weather satellites can monitor weather and cloud patterns covering an entire hemisphere of the Earth with a fresh image of the full disk of the Earth available every 30 min [2]. For example, METEOSAT is a geostationary satellite that provides high temporal resolution, which enables continuous monitoring of clouds and weather conditions. However, geostationary satellites have limited spatial resolution when compared to polar-orbiting satellites due to their high altitude. Another disadvantage is incomplete geographical coverage, since ground stations at latitudes higher than 60° have difficulty receiving signals at low elevations.

Polar-orbiting satellites provide coverage of Earth's polar regions beyond the view of geostationary satellites and fly at altitudes typically between 800 and

S.-Y. Tan, *Meteorological Satellite Systems,* SpringerBriefs in Space Development, 19
DOI 10.1007/978-1-4614-9420-1_3, © The author 2014

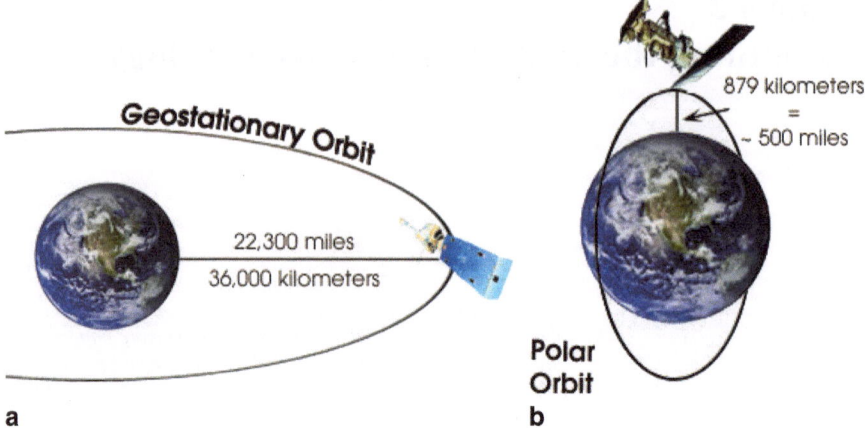

**Fig. 3.1** The two main satellite orbits, **a** geostationary earth orbiting satellites and **b** polar-orbit orbiting satellites. (Courtesy of University of Wisconsin-Madison Space Science and Engineering Center) [1]

1,200 km above Earth in a north to south (or vice versa) path (Fig. 3.1b). Polar-orbiting satellites are in Sun-synchronous orbits, observing Earth with the Sun at a constant angle to the satellites every day throughout the year. The angle relative to the Sun is determined by the satellite equator-crossing time, which is chosen at the time of launch of the satellite [3]. These Sun-synchronous satellites are able to observe almost any place on Earth, including polar regions, and will view every location twice in 24 h with constant general lighting conditions due to the near-constant local solar time. Recording conditions stay constant and scenes from different time periods can be easily compared.

Polar orbits are able to image polar regions more frequently than areas at low latitudes due to the increasing overlap in adjacent swaths as the orbit path comes closer together near the poles. The swaths are usually about 2,600 km wide. Polar-orbiting satellites tend to provide higher spatial resolution images due to their proximity to Earth, as well as regular data collection at consistent times for long-term comparisons.

Satellite orientation and stabilization in space is highly important for solar cells receiving solar radiation and antenna communicating information. Satellites intended for weather and telecommunication purposes usually require to be earth-pointing and it is necessary to control the satellite's position in both east-west and north-south directions. However, once in orbit, a satellite experiences perturbing torques, including gravitational forces, magnetic fields, and solar radiation pressures, which can affect satellite orientation [4].

An attitude and orbit control system maintains a satellite's position and orientation, usually using a set of thrusters that are fired to achieve a desired rate of rotation and to reach a desired position. Satellites can be classified by their techniques used

**Fig. 3.2** TIROS-6 satellite, an example of a spin-stabilized satellite. Note the vidicon lens at bottom left on the satellite. (Courtesy of NASA) [6]

for attitude control, including most typically either (a) spin-stabilized satellites, and (b) three-axis stabilized satellites.

## *Spin Stabilized Satellites*

Spin-stabilized satellites have the motion of one axis relatively fixed by spinning the satellite around the major axis by the gyroscopic effect offering inertial stiffness, much like a spinning top. This prevents the satellite from drifting from its desired orientation and keeps the spacecraft pointed in a certain direction. The spinning spacecraft effectively resists perturbing forces, which tend to be small in space. Spin-stabilization design was applied to most early satellites and typically have a cylindrical shape.

Figure 3.2 shows the TIROS satellite (TIROS-6) as an example of this early design. In this configuration, solar arrays are mounted around the spinning cylindrical drum. One disadvantage of this design is that the satellite is limited in terms of the size of solar arrays that can be used to obtain solar energy, which results in large battery power requirements. Another disadvantage is that instruments or antennae must also perform 'de-spin' maneuvers, so that the satellite antenna can point at the appropriate Earth terminals [5]. Moreover, this design suffers from the fact that instruments on the satellite spend much of their time looking away from Earth and only a small part of their time facing Earth. Finally, since the spin rate of the satellite decays with time, constant readjustment is required. This readjustment can be achieved by firing thrusters tangentially to the circumference of the spinning drum.

## *Three-Axis Stabilized Satellites*

For three-axis stabilized satellites, the body of the satellite remains fixed in space relative to a constant Earth orientation. Thus the movement of the satellite is controlled along all three axes (i.e., yaw, pitch, and roll). Small spinning wheels (called reaction wheels or momentum wheels) rotate to maintain the spacecraft in the desired orientation in relation to Earth and the Sun. Corrective forces are applied by the active control system on the wheels to correct for satellite orbit perturbations. Once satellite sensors detect that the satellite is moving away from the proper orientation, momentum wheels immediately speed up or decelerate to provide the dynamic force required to keep it within a specific range. Some spacecraft also use small propulsion-system thrusters to continually nudge the spacecraft back and forth until the correct position is achieved. Examples of weather satellites belonging to the category of three-axis stabilized systems include GOES-8, GOES-9, and TIROS-N satellites.

In comparison to spin-stabilized satellites, three-axis stabilized satellites can view Earth continuously, but are somewhat more difficult to maintain a perfect relationship or orientation to Earth. They also have an extendible solar array that is unable to provide power when the satellite is in the transfer orbit, while it is still stored inside the satellite. However, three-axis stabilized satellites have potential for more power generation capability with large solar arrays during operation, as well as additional mounting areas available for complex antennae structures.

They are also able to point optical instruments and antennae without having to de-spin them, unlike spin-stabilized satellites. However, spin-stabilized satellites are often simpler in design, lighter in mass, and usually less expensive. Figure 3.3 shows GOES-8, an example of a three-axis stabilization configuration.

## Comparison of Polar-Orbiting and GEO Satellites

Since 71 % of Earth's surface is covered by water, and large regions are sparsely inhabited, polar-orbiting satellite systems produce data that compensate for the deficiencies in conventional observing networks. The polar-orbiting spacecraft offers the advantages of acquiring data from almost all locations on the globe in the course of successive revolutions (e.g., TIROS series and POES series satellites). The primary advantages of this type of satellite are observing daily global cloud cover, accurate quantitative measurements of surface and vertical temperature, and atmospheric water vapor [8].

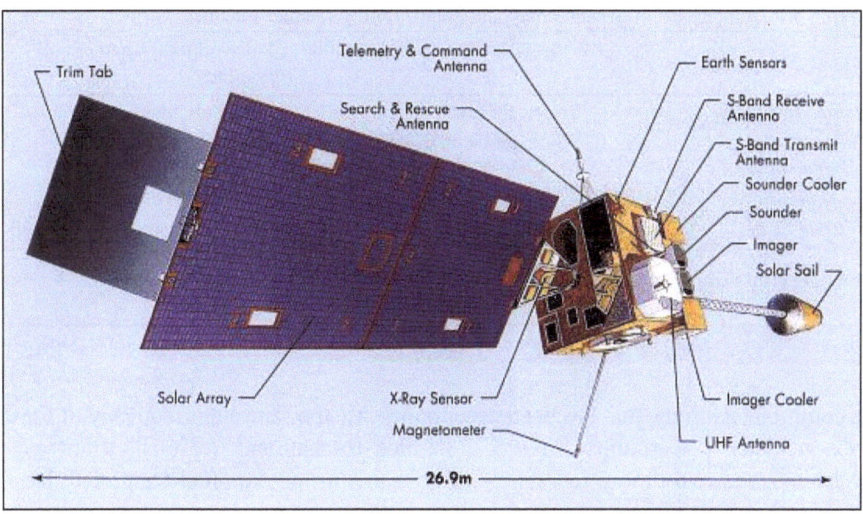

Telemetry & Command Antenna
Trim Tab
Earth Sensors
Search & Rescue Antenna
S-Band Receive Antenna
S-Band Transmit Antenna
Sounder Cooler
Sounder
Imager
Solar Sail
Solar Array
X-Ray Sensor
Magnetometer
Imager Cooler
UHF Antenna
26.9m

**Fig. 3.3** GOES-8 satellite, an example of a three-axis stabilized satellite. (Courtesy of eoPortal) [7]

Moreover, Sun-synchronous polar-orbiting satellites offer the advantages of moderate to high spatial resolution, and nearly constant daily equatorial crossing times. They are able to track weather conditions, atmospheric variables, and cloud cover. However, observations of instant changes from these platforms are spatially and temporally fragmented due to the swath width and revisit times of the orbital ground tracks. A method to mitigate this loss is to fly sensors on twin satellites, so that the sensor can sample every 6 hours from a near-polar orbit. Nevertheless, a good combination of spatial and temporal resolution is still required for adequately sampling and monitoring a phenomenon of interest.

In contrast, geostationary satellites provide a continuous view of weather systems necessary for intensive data analysis, thereby recording the motion, development, and decay of changing atmospheric conditions. Geostationary satellites "hover" continuously over a fixed position on Earth's surface. For example, severe thunderstorms with a lifetime of only a few hours can be successfully detected and recognized in the early stages, making it possible to issue warnings to the general public of the time and area of the storm's maximum impact. Geostationary satellite imagery can also be used to estimate rainfall during thunderstorms, flash floods during hurricanes, and snowfall accumulations in the winter. Due to this capability, geostationary spacecraft has been primarily used for short-term natural disaster warnings.

Table 3.1 summarizes and compares basic features of geostationary and polar orbiting satellites. Given the opposing advantages and disadvantages of polar-orbiting and geostationary satellites, it has been suggested that the strengths of each platform complement each other and could be greatly exploited by merging datasets [9]. Analyzing satellite imagery collected from multiple sensors and/or multiple platforms is

**Table 3.1** Comparison of geostationary and polar-orbiting weather satellites

|  | Geostationary weather satellites | Polar-orbiting weather satellites |
| --- | --- | --- |
| *Image resolution* | Relatively low | Relatively high |
| *Scan coverage* | Whole globe disk | Limited areas |
| *Altitude of orbit* | About 36,000 km (22,370 miles) above the equator | About 800–1,200 km (497–746 miles) |
| *Movement of satellite* | Orbit in synchronization with the Earth | Circling the Earth in a roughly north-south orbit |
| *Frequency of image capture* | Continuous viewing of one location | One to two times a day for the same place |

a common technique that has been used to increase the sampling frequency of Earth observations. For example, GOES is applied for national, regional, short-range warnings and real-time forecasts, while POES is usually applied for global, long-term forecasting and environmental monitoring.

Producing image composites by combining GEOS and PEOS observations is challenging due to the need to deal with differences in calibration, viewing geometry, and temporal offsets from a variety of satellites. Temporally, two major factors to contend with are the timeliness of the data and time interval between composite images. Advances in multi-sensor data fusion techniques optimize the use of current weather satellite systems and integrate data for global weather monitoring and improved forecasts. Combined together, both types of satellites constitute a truly global meteorological network, evolving into an integrated environmental observing system with capabilities to observe, assess, and predict the total Earth system, including atmosphere, ocean, land, and space environments.

## Evolution of Sensing Techniques

In the early 1960s, television cameras were on-loaded onto meteorological satellites to view and monitor Earth from low orbit altitudes of approximately 720 km. Pictures taken by a 10 mm lens at a speed of 5 frames per set achieved a spatial resolution of 0.15 km under full sunlight with 25 % contrast [10]. These parameters permitted a transverse strip approximately 550 km wide to be photographed when a satellite orbited Earth.

TIROS-1 marked the world's first weather satellite to test the experimental television techniques (Fig. 3.4). It carried two 6-inch long television cameras on-board, which were rugged and lightweight devices weighing only about 2 kg. One of the cameras had a low resolution wide-angle lens providing views of approximately 1,200 km on a side, while a narrow angle camera had a higher resolution telephoto lens with a view that was about 130 km on a side [11]. A magnetic tape recorder for each camera was supplied for storing photographs when the satellite was beyond the

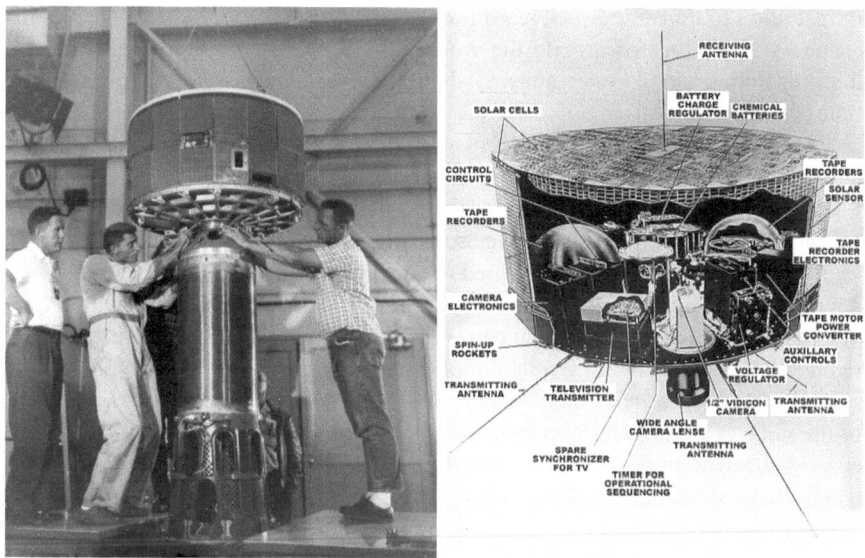

**Fig. 3.4** TIROS-1 was the first successful dedicated weather satellite (*left*) and was equipped with two cameras (*right*). (Courtesy of NASA and NOAA) [12, 13]

range of a ground station, containing 122 m of tape and recording up to 32 pictures. When within range of a ground station, the cameras took pictures every 10–30 s.

In addition to vidicon camera systems for daytime visible imaging, TIROS satellites also carried passive infrared radiometers for sensing during both day and night conditions. TIROS-8, launched in December 1963, carried a 1.27-cm vidicon camera and a 2.54-cm automatic picture transmission (APT) camera. The latter utilized a very slow-scan vidicon compared to the former and was designed for real-time picture transmission of cloud cover conditions to be received by fairly simple and inexpensive ground stations anywhere in the world. Whereas the TIROS vidicon camera required 2 s to scan its 500-television-line image, the APT camera required 200 s for read-out of its 800-television-line image [14]. Over 1,600 pictures were obtained in the 3½ weeks of mission lifetime. The APT cameras were subsequently used on Nimbus-1 and -2, then on the operational ESSA series and the initial NOAA satellites.

In the ESSA series (also called the TIROS operational System – TOS series), satellites from ESSA-3 began to apply a modified Nimbus camera, the Advanced Vidicon Camera System (AVCS). This system combined APT transmission and on-board data storage for the collection of daytime imagery (visible and thermal infrared) and night-time (thermal infrared only). The AVCS consisted of three cameras, a tape recorder, and an S-band transmitter that produced a three-segment composite picture. It provided higher resolution imagery and a larger picture area than the 1.27 cm vidicon for TIROS. It also yielded a linear resolution of better than 1 km at nadir from an altitude of 800 km [15]. The 2 kHz bandwidth of the system enabled the satellites to transmit direct, real-time television pictures to the inexpensive APT ground stations located around the world. AVCS provided more near-global and high-resolution cloud cover imagery than had ever been assembled prior to that time.

ESSA was followed by the Improved TOS (ITOS, or TIROS-M series), which allowed the global coverage and APT services to be combined on one satellite. Launched in January 1970, ITOS-1 was the first of an operational series of three-axis stabilized satellites. This allowed scanning radiometers (SR) to be flown on the satellites and provided routine infrared window coverage both day and night. These devices were used to further evaluate and develop global sea-surface temperature mapping, as well as other thermal patterns, which led to significant contributions to radiation budget and space weather applications.

Only about three years later, the launch of NOAA-2 (or ITOS-D) marked the end of the vidicon era and the start of the era of multi-channel high-resolution radiometers, which was an improvement over the SR. The Very High Resolution Radiometer (VHRR), a calibrated scanning radiometer, was installed on the TIROS-M series initially only to capture information from the visible and infrared window. As part of international weather data exchange, NOAA introduced the direct reception of VHRR data at no charge to ground stations built by an increasing number of users, beginning in 1972. ITOS-1 and NOAA-1 were transitional satellites of the ITOS series. The NOAA-2 through NOAA-5 satellites that were launched in 1972–1976, carried the VHRR instrument.

The next milestone of the development of imaging technologies occurred on October 1978 with the launch of TIROS-N. This satellite included an across-track scanning instrument, in the form of a four spectral channel radiometer called the Advanced VHRR (AVHRR). This advanced radiometer covered the visible, near-infrared, mid-wave infrared, and thermal spectral channels. All four channels had a spatial resolution of 1.1 km. The infrared window channels had a precise thermal resolution of 0.12 K at 300 K. The AVHRR installed on later satellites were applied for global daytime and night-time sea surface temperature determination, heat budget components estimation, cloud delineation, and snow and sea ice identification. The AVHRR subsequently improved to a five and six spectral channel system, with all channels providing imagery at 1.09 km resolution at nadir. The data has also been used for a variety of other studies, including mapping marine oil pollution, monitoring volcanic eruptions, and assessing vegetation vigor on an international scale. Information about the wavelength and typical applications of the six bands of AVHRR are listed in Table 3.2.

A second primary sensor on-board the TIROS-N was a vertical sounder, referred to as the Vertical Atmospheric Sounder (VAS). This is a type of radiometer that measures infrared or microwave radiation. This device provided vertical profiles of temperature, pressures, water vapor, and critical trace gases (e.g. carbon dioxide or ozone) in Earth's atmosphere. Apart from the VHRR, NOAA-2 (launched in 1972), also allowed operational thermodynamic soundings with the Vertical Temperature Profile Radiometer (VTPR). This instrument was an 8-channel TIR/FIR filter radiometer that scanned perpendicular to the satellite motion with a horizontal resolution of $55 \times 57$ km at the nadir and $67 \times 91$ km at the edges of the radiometer's scan [17].

**Table 3.2** AVHRR/3 sensor channel characteristics and typical applications. (Reproduced from NOAA) [16]

| Channel # | Wavelength (μm) | Applications |
|---|---|---|
| 1 | 0.58–0.68 | Daytime cloud and surface mapping |
| 2 | 0.725–1.00 | Land-water boundaries |
| 3A | 1.58–1.64 | Snow and ice detection |
| 3B | 3.55–3.93 | Night cloud mapping, sea surface temperature |
| 4 | 10.30–11.30 | Night cloud mapping, sea surface temperature |
| 5 | 11.50–12.50 | Sea surface temperature |

In the same period, it was recognized that the optimum temperature profiles would be obtained by taking advantage of the unique characteristics offered by taking soundings at the 4.3 μm, 15 μm, and 0.5 cm wavelength regions. Consequently, the High Resolution Infrared Radiation Sounder (HIRS) experiment of Nimbus-6 (launched in 1975) was designed to accommodate channels in both the 4.3 and 15 μm infrared regions. The 7-channel HIRS was about $25 \times 25$ km in spatial resolution. It was also designed to measure scene radiance to permit the calculation of the vertical temperature profile from Earth's surface all the way up to about 40 km in altitude. This was subsequently complemented by the 0.5-cm microwave wavelength channels of a Scanning Microwave Spectrometer (SCAMS), which provided nearly full Earth coverage every 12 hours. The combination of instruments on-board the Nimbus-6 satellite provided improved sounding capabilities as compared to the infrared sounders alone.

The TIROS-N/NOAA satellites series that began in 1978 not only carried the AVHRR but also an atmospheric sounding system called the TOVS-TIROS Operational Vertical Sounder (TOVS). This sounder provided vertical profiles of temperature and water vapor from Earth's surface to the top of the atmosphere. A solar proton monitor detected the arrival of energetic particles for applications in solar storm prediction [18]. The TOVS incorporated a suite of instruments, including a HIRS, the Microwave Sounding Unit (MSU), and the Stratospheric Sounding Unit (SSU). An operational Solar Backscatter Ultraviolet Radiometer (SBUV/2) used to monitor the distribution of ozone in the atmosphere was flown on NOAA-9, -11, -13, and -14 satellites. Since NOAA-15 (launched in 1998), all sounding units have greatly improved, including the deployment of the Advanced TOVS (ATOVS) system for NOAA-15 and subsequent satellites. The ATOVS used the HIRS and AMSU-A to generate atmospheric profiles, while the AVHRR instrument was used for cloud detection. The ATOVS from NOAA-15 (launched in 1998) generated about 300,000 retrievals every 24 h with a 60-km spatial resolution. It was the first in a series of five satellites with improved imaging and sounding capabilities.

NOAA-15 also first carried the AVHRR/3 sensor (Table 3.2), which is an across track scanner that senses Earth's outgoing radiation in six channels. Three channels operate in the visible-near infrared region and three channels in the thermal infrared (TIR), with a spatial resolution of 1 km at nadir. The data from the six channels, after processing, permits multi-spectral analysis for more precisely defining hydrologic, oceanographic, and meteorological parameters. AVHRR/3 data has been used widely for measuring temperature and various environmental parameters of land, water, sea surfaces, and cloud cover [19].

## Data Synthesis and International Data Exchange

Researchers have increasingly applied data fusion methods to integrate and display data from multiple space-borne sensors in order to analyze complementary remote sensing imagery and thus to fulfill research objectives. For example, a model yielding the probability of rainfall was established based on GOES visible/infrared imagery and simultaneous radar data for eastern Canada in 1980 [20].

A similar approach has been applied for daily evapo-transpiration mapping implemented by the Atmosphere-Land Exchange Inverse (ALEXI) model, which is a multi-sensor TIR approach [21]. This model was developed for estimating and mapping surface fluxes and surface moisture variables at the regional scale. The TIR imagery is derived from any type of GOES data, depending on the resolution required by a given application. The ALEXI model has potential for global applications by integrating data from multiple geostationary meteorological satellites systems, such as the American GOES, the European METEOSAT satellites, the Chinese Feng-yun 2B series, and the Japanese GMS. Observation covering scales of these international GEO satellites is shown in Fig. 3.5.

Synthetic application of multi-sensor data can be found in many research studies and applications. For example, the International Satellite Cloud Climatology Project (ISCCP), the first project of the World Climate Research Program (WCRP), established in 1982, has the goal of examining global distribution of cloud radiative properties. By collecting and analyzing satellite radiance measurements, global cloud distribution, and other data, it is possible for temporal (diurnal, seasonal, and yearly) variations to be inferred (see Fig. 3.6). Data applied in this project have been sought from five geostationary meteorological satellites, including GOES-E, GOES-W, GMS, INSAT (Indian National Satellite System), and METEOSAT, as well as at least two polar-orbiting satellites (NOAA/TIROS-N and METEOR-class satellites). The primary datasets have been derived from two standard visible (0.6 μm) and infrared (11 μm) channels from all of these satellites [23]. ISCCP cloud data are used to determine cloud effects on Earth's radiation balance. This has greatly enhanced understanding of cloud formation processes and the global water cycle.

Another example of international satellite data exchange is the International Space Environment Service (ISES). This global service facilitates near-real-time international monitoring and prediction of the space environment. ISES comprises a network of globally distributed Regional Warning Centers. A permanent service of the Federation of Astronomical and Geophysical Data Analysis Services (FAGS), the ISES has an important role in coordinating the exchange of data between organizations around the world, who are involved in forecasting solar terrestrial conditions and reducing the impact of space weather on activities of human interest [25].

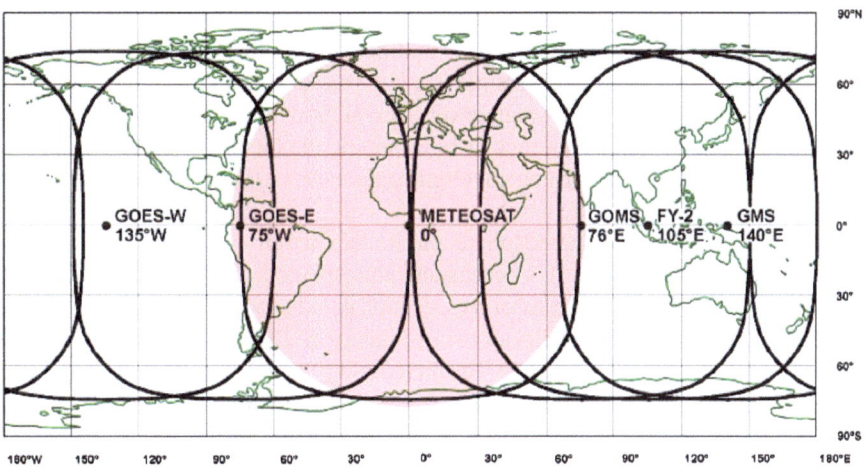

**Fig. 3.5** Global observation coverage by international geostationary meteorological satellites. (Courtesy of Seiz 2010) [22]

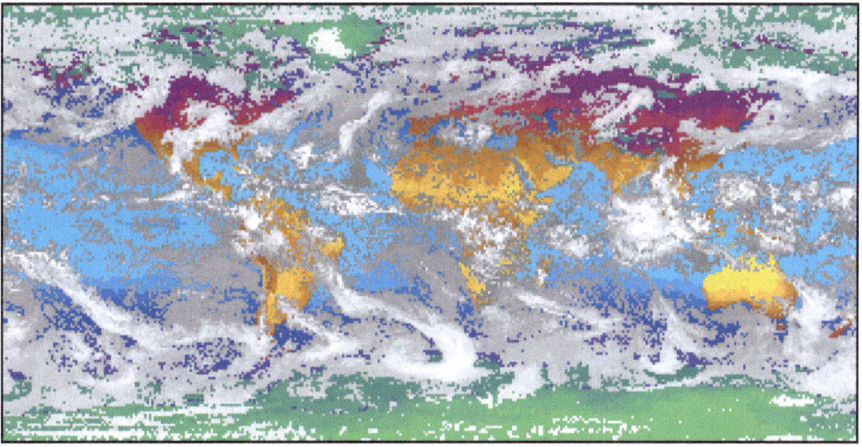

**Fig. 3.6** NASA ISCCP data product image of global cloud coverage. (Courtesy of NASA ISCCP) [24]

The data exchanged are highly varied in nature and in format, ranging from simple forecasts or coded information up to more complicated information, such as satellite imagery.

An important strength of the ISES data exchange system is the network of Regional Warning Centers and access to data from unique instrumentation available from the scientific community in each center's region. Exchange via the ISES enables such data sets to be available to the wider international scientific and user community. A data exchange schedule operates with each center providing and

relaying data to other centers. A center in Boulder, Colorado (USA), plays a special role as a "world warning agency" and acts as a hub for data exchange and
space weather forecasts. The ISES plays a key role in international data exchange
based on cooperation in space weather services. Within the European Space Agency
(ESA), data is exchanged on a regular basis among the member states.

The U.N.'s World Meteorological Organization (WMO) serves as the global au-
thoritative voice on the state and behavior of Earth's atmosphere [26]. The WMO
is thus concerned with the atmosphere and its interaction with the oceans, the impacts on the climate that Earth's oceans and atmosphere actually produces, and the
resulting distribution of water resources. Established in 1950, the WMO comprises
of 191 member states and territories (as of January 1, 2013). The WMO is thus the
United Nations' specialized agency for meteorology (weather and climate), operational hydrology, and related geophysical sciences. The WMO promotes cooperation between member states for making meteorological observations. It carries out
its mission by facilitating the free and unrestricted exchange of data and information, processing and standardization of related data, and assisting with technological
transfer, training, and research. The WMO and its programs for weather monitoring
and warning are further discussed in Chap. 7.

Since weather, climate, and the weather cycle know no national boundaries, international cooperation at a global scale is essential. With the rapid new development of instrumentation to quantify meteorological phenomenon and a growing
constellation of weather satellites collecting real-time data in orbit, it is clear that
there is much more data being collected and thus much more to be processed and
stored. With better and easier ways to obtain and store information, there also comes
a higher demand for ways of managing and analyzing all the data. This is vital not
only to public weather services and applications in agriculture, aviation, shipping,
environment, and water issues, but to many other important areas as well.

Proper processing and use of key data has incalculable benefits for humankind's
well-being, providing vital information for advance warnings that save lives and reduce damage to property and increasingly to help preserve the environment. Global
weather data exchange is essential for climate monitoring and for better understanding the world's weather system to help prepare and protect human populations from
adverse and extreme weather events. The effective use of this data might be perhaps
tomorrow, next week, next year, or perhaps decades in the future.

# References

1. University of Wisconsin-Madison Space Science and Engineering Center. Satellite applications for geoscience education. http://cimss.ssec.wisc.edu/sage/remote_sensing/lesson1/con-
   cepts.html. Accessed 12 Aug 2013
2. Hillger, D.W.: Geostationary weather satellites. Topical Time, 48(2), 41–42 (1997)
3. Hillger, D.W.: Polar-orbiting weather satellites. Topical Time, 48(4), 33–36 (1997)
4. Maini, A.K., Agrawal, V.: Satellite technology: principles and applications. Wiley (2011)

5.  Darrin, A., O'Leary, B.L.: Handbook of space engineering, archaeology, and heritage. CRC Press (2010)
6.  NASA: http://history.nasa.gov/SP-168/appendix.htm. Accessed 12 Aug 2013
7.  eoPortal:  https://directory.eoportal.org/web/eoportal/satellite-missions/g/goes-2nd-generation. Accessed 12 Aug 2013
8.  Hinsman, D.E.: The role of geostationary satellites in WMO's Global Observing System. GOES Users' Conference held in Boulder, Colorado, May 22–24, 2001, 34–35 (2001)
9.  Eva, H., Lambin, E.F.: Remote sensing of biomass burning in tropical regions: Sampling issues and multisensory approach. Remote Sensing Environment, 64, 292–315 (1998)
10. Greenfield, S.M.: The initial conceptualization and design of a meteorological satellites. Meteorological satellites past, present, and future, 3–6 (1982)
11. NOAA Satellite and information service: http://noaasis.noaa.gov/NOAASIS/ml/40yearsa. html. Accessed 12 Aug 2013
12. NASA: http://www.nasa.gov/centers/goddard/about/didyouknow_prt.htm. Accessed 12 Aug 2013
13. NOAA: http://noaasis.noaa.gov/NOAASIS/ml/40yearsa.html. Accessed 12 Aug 2013
14. Sclinapf, A.: The TIROS Decade. IEEE, 53–59 (1969)
15. Kramer, H.J.: Observation of the Earth and Its environment: survey of missions and sensors, 4th edn., Springer (2001)
16. NOAA: http://noaasis.noaa.gov/NOAASIS/ml/avhrr.html. Accessed 12 Aug 2013
17. NOAA/NESDIS/National Climatic Adata Center: http://www.ncdc.noaa.gov/oa/rsad/vtpr. html. Accessed 12 Aug 2013
18. Manna, A.J.: 25 Years of TIROS Satellite. Bulletin of American Meteorological Society, 66(4), 421–423 (1985)
19. EUMETSAT: http://www.eumetsat.int/Home/Main/Satellites/Metop/Instruments/SP_2010053153142514. Accessed 12 Aug 2013
20. Bellon, A., Lovejoy, S., and Austin, G.L.: Combining satellites and radar data for the short-range forecasting of precipitation. Monthly Weather Review, 108, 1554–1566 (1980)
21. Anderson, M.C., Normal, J.M., Diak, G.R., Kustas, W.P., Mecikalski, J.R.: A two-source time-integrated model for estimating surface fluxes using thermal infrared remote sensing. Remote Sensing of Environment, 60, 195–216 (1997)
22. Seiz, G.: Satellite meteorology. http://Earth.eo.esa.int/eosummerschool/envschool_2010/GS3.pdf (2010). Accessed 12 Aug 2013
23. NASA ISCCP: http://isccp.giss.nasa.gov. Accessed 12 Aug 2013
24. NASA ISCCP: http://isccp.giss.nasa.gov/ISCCP.html. Accessed 12 Aug 2013
25. ISES: http://www.ises-spaceweather.org. Accessed 12 Aug 2013
26. WMO: http://www.wmo.int. Accessed 12 Aug 2013

# Chapter 4
# U.S. Meteorological Satellites

> *Scientists have studied the atmosphere for many decades, but its problems continue to defy us. ...Here, new scientific tools have become available. With modern computers, rockets and satellites, the time is ripe to harness a variety of disciplines for a concerted attack.*
>
> —President John F. Kennedy (U. N. General Assembly, September 25, 1961)

This chapter focuses on the evolution of the U.S. meteorological satellite program. We will explore the historical and complementary roles of NASA, NOAA, and the Department of Defense in developing technologies and analyses, as well as the use of data from meteorological satellites. NASA and NOAA have a long history of cooperation in developing spacecraft. An agreement between the two agencies, originally signed in 1973, gives NOAA responsibility for operating the environmental systems and requires NASA to fund development of new systems and to manage research satellite systems.

Through the years, NOAA in cooperation with NASA has successfully launched GOES and POES. Such systems have evolved to become increasingly advanced, with higher spatial and temporal resolution imagery, operational soundings, atmospheric temperature, and moisture data acquisition in all weather situations. This chapter explores how the U.S. meteorological satellite program is organized and how government agencies cooperate and support weather satellite development, operation, and output data for various Earth applications.

## The Development Role of NASA

The history of the U.S. space program can be traced back to initial rocket, sensor, and satellite development projects by the U.S. DOD that began soon after World War II. V-2 rockets were launched with cameras mounted in the nose, which took photos of the Earth. Cameras were parachuted and recovered on the ground. With little theoretical background and experience available to guide these projects, development teams learned their space trade on the job, which often resulted from lessons learned from catastrophic failures [1]. In initial testing, rockets carrying cameras to determine the attitude of the nose cone in space and other instruments

S.-Y. Tan, *Meteorological Satellite Systems,* SpringerBriefs in Space Development, DOI 10.1007/978-1-4614-9420-1_4, © The author 2014

developed by DOD were able to view and film the Earth below, showing interesting cloud patterns as they reached higher altitudes. These first ventures from rockets were practical and efficient ways of collecting environmental information that was critical to advancing American space efforts. Images from these cameras offered investigators synoptic views of weather patterns, providing a new platform from which to observe Earth from space and near space. However, data and measurements were collected from salvaged recorders or radio transmissions. These initial experiments demonstrated the need for satellites in order to observe weather conditions around the globe and on a continuous basis.

In the mid-1950s, the U.S. Weather Bureau (USWB), under the Department of Commerce (DOC), initiated the Weather Satellite Program. In March 1958, the chief of the USWB, Dr. F. W. Reichelderfer, established a special unit called the Meteorological Satellite Research Unit, comprised of three sections: Meteorological Satellite Instrument Development, Meteorological Satellite Research Unit, and Two Field Meteorological Satellite Telemetering Units [2]. On July 29, 1958, Eisenhower signed the National Aeronautics and Space Act, establishing NASA. This agency absorbed the 46-year old National Advisory Committee for Aeronautics (NACA). The bill would change NACA's former role of research to include large-scale development, management, and operations.

In September 1958, a memo was sent to the USWB units indicating the formation of NASA, indicating that the weather bureau would be designed as their meteorological agent, providing the meteorological instrumentation, data reduction, and analysis of observations taken by satellites after the International Geophysical Year (IGY) series finished [1]. As the small USWB group expanded with more people recruited, its name changed to the Meteorological Satellite Laboratory (MSL), marking the beginning of NOAA's major satellite program. The MSL further evolved into today's NESDIS (National Environmental Satellite, Data, and Information Service), which provides timely access to global environmental data from satellites and information services.

NASA was basically created due to pressures related to national defense. After World War II, the United States and the Soviet Union were engaged in the Cold War, with both major powers struggling for global influence. During this period, space exploration emerged as a significant area of competition, with both nations focused on attaining firsts in space technologies, which were symbolic of technological and ideological superiority. The space race began with the Soviet Union's launch of the Sputnik-1 artificial satellite on October 4, 1957. Alarmed by the perceived threat to national security and technological leadership, NASA was founded as a new federal agency based mainly on the NACA. NASA began to conduct space missions within months of its creation, including remote sensing meteorological and Earth science missions. Currently, NASA invests around $ 1.5 billion per year on developing satellites to observe and monitor Earth out of an overall budget of about $ 18 billion [3].

Furthermore, most NASA research satellites and instruments have provided advanced measurement and satellite technologies that contribute to operational monitoring system programs and satellite systems at NOAA and the DOD. Operation of weather satellites was transferred to NOAA in the DOC, a successor agency of the ESSA, at the beginning of the 1960s. However, NASA remains the lead agency responsible for launching and collecting Earth observation data from space and for sponsoring fundamental research and analysis of those data.

## The Operational Role of NOAA

On October 3, 1970, NOAA was formed under the DOC. This was proposed by Richard Nixon to serve a national need "...for better protection of life and property from natural hazards...for a better understanding of the total environment...[and] for exploration and development leading to the intelligent use of our marine resources." [4] Specifically, NOAA's mission is to describe and predict Earth's environment so as to safeguard lives and property and to contribute to national economics and environmental health. In order to support and fulfill this mission, NOAA gathers environmental data around the world and from space [3].

Operating the country's environmental satellite program, NOAA is responsible for developing and applying space-based Earth remote sensing and cloud imagery for NOAA's National Weather Service (NWS) forecasts. The NWS is responsible for weather warning services, providing forecasts, and other products for the purposes of protection, safety, and general information. Within NOAA, the NESDIS office operates the satellites and manages the processing distribution, as well as the archiving of the data. It distributes on a global basis more than 3.5 billion vital bits of data and images to forecasters daily [1]. Data is used for various applications, including oceanography, agriculture, forest fire detection, volcanic ash monitoring, monitoring atmospheric ozone, and storm prediction. Over time, NOAA's satellites have evolved from serving solely as weather satellites to environmental satellites for broad applications, thus enhancing our understanding of climate variability and society's ability to plan and respond. Its role with regard to monitoring space weather has also expanded over time.

NOAA's operational activities, such as weather prediction, routinely and reliably generate specific services and products that meet predefined accuracy, timeliness, and scope/format requirements. In addition to providing cloud images for daily television weather forecasts, the information is also disseminated or made available to a variety of users in the public, private, and academic sectors. Operational systems must be developed to withstand risk and operate without interruption to provide measurements with predefined accuracy and timeliness requirements. NOAA represents the U.S. investment in satellite technologies and infrastructure to routinely collect data and transmit operational information about Earth's environment (weather, climate, oceans) [5].

The NOAA satellite constellation consists of complementary operational environmental satellites, including GOES and POES. Both types of satellites are critical to the global weather network. As noted above, the NESDIS processes and distributes more than 3.5 billion vital bits of data and images to meteorologists and forecasters globally, and it does so under demanding standards of timeliness and quality [6]. Further in doing so, it is constantly working to accurately combine polar and geostationary satellite data. This mission is being achieved by virtue of greatly upgraded data processing and analysis installations, enhanced ground facilities, and data sharing agreements with military weather services. The primary customer of these data remains the NWS, which adopts satellite data to perform weather forecasts for the public, television, radio, and weather advisory services. Satellite information is also shared with various federal agencies, other countries (e.g., Japan, Russia, European Space Agency members, and the United Kingdom Meteorological Office), and the private sector.

## DOD's Mission Objectives for Meteorological Satellites

With the establishment of NASA in 1958 as the U.S. civil space agency, the DOD retained responsibility for space matters bearing on national security interests. Previous military sponsored experimental satellite systems for weather observations that were developed in the 1950s were transferred to NASA. This included the Joint Astrophysics Nascent Universe Satellite (JANUS) project, a classified reconnaissance project for measuring star formation, which became the predecessor to TIROS-1.

Both NASA and DOD continued research and development efforts for improving Earth-observation capabilities, including data interpretation and improving onboard sensor technologies. In the 1960s, it became clear that the operation of certain highly classified military programs could be significantly enhanced if cloud cover over the Eurasian continent were known accurately and on a daily basis. This need could not be met by the National Meteorological Satellite Program. Under the direction of DOD and the Undersecretary of the Air Force, a long-term satellite program for monitoring the meteorological, oceanographic, and solar-geophysical environment of Earth to specifically support DOD operations was approved on August 21, 1961. Originally named Program 35, the name was subsequently changed to Program 417, or the Defense Satellite Applications Program (DSAP). This has since been superseded by the DMSP, which is a total system involving sensors, data communications, and ground processing equipment. In its 50-year history, 41 DMSP satellites have been successfully launched.

According to the DMSP User's Guide [7] released in 1974, the mission of the program was to (a) provide globally recorded visual and infrared cloud cover and other specialized environmental data to the Air Force Global Weather Central (AFGWC); (b) to provide real-time direct readout of local area environmental data to mobile receiving terminals at key locations throughout the world; and (c) to continue the advancement of environmental satellite technology to meet DOD requirements. The program was constrained with strict security provisions, a hard funds ceiling, a program schedule ($10 million with 9 months to first launch), and fixed price contracts on equipment. Strict budgetary constraints were necessary, since a substantial budget had been allocated for the NOAA satellite program and a second military/defense satellite program would be difficult to justify to the public.

Currently, the DOD invests nearly $ 1 billion each year in operational, research and development (R&D) related to environmental data acquisition and the use of these data in support of national security [3]. Specifically, the U.S. Air Force (USAF) has applied meteorological products to a full range of activities, including mission planning (e.g., force prepositioning), operational support (e.g., air tasking orders), and tactics (e.g., target selection). It has maintained an operational constellation of two near-polar, Sun-synchronous satellites for military purposes since the first launches of Block 1 DMSP satellites in 1962. Although originally limited to "need-to-know" personnel, DMSP data was declassified in December 1972 by President Nixon and made part of the public domain, available to the civil and scientific communities.

## Tri-Agency Partnership and Coordination

In the late 1960s and early 1970s, the three agencies (NASA, DOD, and NOAA) developed a symbiotic and productive relationship. This consequently led to a period of phenomenal discovery and development in remote sensing, as NOAA was able to build off of environmental satellite technology previously developed by NASA and the DOD. Meanwhile, NASA and the DOD obtained insights from NOAA regarding the conduct of daily satellite operations, data processing, timely delivery of products, and application of these data [1].

Although tremendous growth in the use of weather satellites for both civilian and military applications has been witnessed in the last few decades, satellites for civilian and military usage have developed largely as separate systems. In fact, since the early 1970s, proposals have been made to merge civilian and military meteorological systems, especially to save on development and operational costs [8]. During this time period, NOAA and its predecessor agencies within the Department of Commerce reimbursed NASA and the DOD for the personnel and other costs they incurred with the purpose of meeting NOAA's space mission. General and specific agreements among the DOD, NOAA, and NASA governed and regulated the relationship, responsibilities, and costs of the support. The specifics of these past agreements are not always in the public domain.

Under a 1994 presidential directive by Bill Clinton, a single tri-agency organization between the DOD, NOAA, and NASA was formed to replace the separate systems and to produce a single national space-based system for environmental monitoring. The Integrated Program Office (IPO) within NOAA was established in October 1994 as a result of the signing of a tri-agency Memorandum of Agreement (MOA). Hence, a National Polar-orbiting Operational Environmental Satellite System (NPOESS) was named to converge the parallel polar-orbiting weather satellite programs of the DOD and DOC, which was largely based on the previous DMSP and POESS program [9].

The goal of this tri-agency was to coordinate the development of the NPOESS satellite partnership that would ultimately meet the needs of all three agencies, while combining command and control operations of military and civil systems. Existing U.S. polar-orbiting satellite systems would be combined under a single national program to collect and disseminate Earth observation data on weather, atmosphere, oceans, land, and near-space environment, with a mandate extending to 2018.

As part of this tri-agency convergence effort, a significant milestone was achieved when NOAA assumed control over DMSP satellites and operational control from the USAF in 1998 [10]. Satellite command was provided by a joint-operational team at NOAA's Satellite Operations Control Center (SOCC). DMSP data were also distributed through NOAA, which was responsible for specifying the performance of the systems needed to satisfy these requirements, obtaining the funds necessary to build and launch the satellites, and building and operating the ground segments of the system. The control and maintenance of the satellites were transferred to NOAA in order to reduce costs, while funding for DMSP continued to be provided by the DOD and the USAF. However, on February 1, 2010, the Executive Office of the President dissolved the next generation polar-orbiting system, NPOESS program. Instead, two separate development programs were announced aimed at serving military and civilian users [11].

## Polar-Orbiting Meteorological Satellites

In April 1959, responsibility for the TIROS research and development program was transferred from the DOD to NASA. TIROS-1, the first satellite capable of monitoring Earth and helping weather forecasts was launched on April 1, 1960 (Fig. 4.1). As an experimental spacecraft, TIROS-1 in its 78-day operating period sent back nearly 23,000 pictures of Earth and its ever-shifting cloud cover from an altitude of about 700 km [12]. This satellite was equipped with both wide- and narrow-angle television cameras, one low-resolution and one high-resolution. However, these two cameras relied on reflected solar radiation, capable of operating only during the daytime.

To assist in overcoming these limitations, as well as for other purposes, TIROS-2 launched on November 23, 1960, carrying several radiometers in addition to two of the same television cameras used on TIROS-1. One of these radiometers measured the energy emitted by Earth in the water vapor window with wavelength interval from 8 to 12 μm. The TIROS experimental system continuously launched ten satellites successfully between 1960 and 1965. It provided early warning of severe tropical storms, hurricanes, and typhoons.

From TIROS-8, launched on December 23, 1963, the Automatic Picture Transmission (APT) camera system and ground stations were tested. The APT camera utilizes a very slow-scan vidicon compared to the television camera [14]. By virtue of the 2 kHz bandwidth of the APT system, TIROS-8 offered the capability of transmitting direct, real-time television pictures to a series of relatively inexpensive

**Fig. 4.1** Launch of TIROS-1
on April 1, 1960. (Courtesy
of NOAA) [13]

APT ground stations located around the world. This technique provided an instantaneous view of local area weather from a satellite passing overhead. In accordance with the NASA-NOAA's USWB agreement, TIROS-9, the first weather satellite to be launched into Sun-synchronous orbit, was a NASA-financed, modified TIROS satellite, orbited to test the new configuration that would eventually be incorporated in the operational series. This configuration maintained the spacecraft spin axis in an alignment normal to the orbital plane and tangent to Earth's surface using the magnetic attitude control system [15].

Before the launch of TIROS-1, limitations (i.e., limited coverage of Earth) caused by its inclined orbit and spin stabilization were recognized. Hence, the design of the Nimbus satellite series began in late 1959 specifically to address these shortfalls with stabilized, Earth-oriented platforms. The Nimbus series first launched in 1965 was designed to test advanced meteorological sensor systems and for collecting atmospheric science data (Fig. 4.2). Its objectives included achieving (a) a near-polar orbit to allow observation of the entire Earth, and (b) Earth stabilization so that the cameras and other sensors could always point toward Earth. Instruments carried on Nimbus included microwave radiometers, atmospheric sounders, ozone mappers, CZCS, and infrared radiometers.

Nimbus-1 and -2 first began to apply an Advanced Vidicon Camera System (AVCS) for recording and storing remote cloud cover pictures. They also carried an

**Fig. 4.2** Nimbus B satellite (*left*). Vibration testing of a Nimbus weather satellite (*right*). (Courtesy of NASA) [16, 17]

APT camera to provide real-time cloud cover pictures and a High-Resolution Infrared Radiometer (HRIR) to complement the daytime TV coverage and to measure nighttime radiative temperatures of cloud tops and surface terrain. In the later five Nimbus satellites, a range of other sensors were tested that were capable of observing different meteorological factors. NOAA scientists made significant contributions to the development of the instrumentation mounted on Nimbus satellites. For example, the satellite infrared spectrometer for measuring atmospheric temperature launched on the Nimbus-3 research satellite was developed in the ESSA MSL.

The early TIROS and Nimbus spacecraft provided the feasibility required for an operational system of meteorological satellites. Its instrument types and orbital configurations opened the door for the development of more sophisticated weather observation satellites [1]. In February 1966, the world's first operational weather satellite system, TIROS Operational System (TOS), was achieved with the successful orbiting of the ESSA-1 and −2 satellites, which were in a near polar, Sun-synchronous orbit. The TOS was sponsored by the DOC, while managed and operated by ESSA's National Environmental Satellite Center (NESC), under the technical direction of the NASA Goddard Space Flight Center (GSFC). The TOS expanded the basic TIROS system and was able to observe Earth's cloud cover routinely on a daily routine. The system provided cloud formation photography to the USWB's National Meteorological Center (NMC) for weather forecasts. In the later TOS satellites, a modified Nimbus camera, the AVCS with higher spatial resolution, was mounted to replace the previous system. Two polar-orbiting TOS satellites, as a pair, were operating at all times to insure uninterrupted daily global photo coverage that met the full operational objectives of the system.

In parallel efforts, the DOD initiated the DMSP, near-polar orbiting satellites providing the military with important environmental information. The main objective of the DMSP by the USAF was to develop a dedicated meteorological system

**Fig. 4.3** The first DMSP launch on May 23, 1962, (*left*) and the DMSP Block-4A satellite (*right*). (Courtesy of National Reconnaissance Office) [18]

that focused on cloud-cover photography to support national reconnaissance collection. DMSP Block I began with five launch attempts on Scout launch vehicles during 1962 and 1963, but four attempts failed. DMSP Block II and Block III satellites launched on Thor Burner I vehicles began to provide weather data for tactical applications in Southeast Asia. Development of more capable and more complex satellites also came to fruition with DMSP Block 4 satellites, seven of which were launched from 1966 to 1969 (Fig. 4.3; [8]).

For civilian weather satellite development, during the second decade of TIROS in the 1970s, the ITOS series witnessed tremendous changes in comparison to the first generation. The first prototype spacecraft of the series was TIROS-M, which achieved a successful orbit on January 23, 1970, and was a joint project of NASA and ESSA. Additional polar-orbiting ITOS meteorological satellites were launched over the next several years by NASA and were financed and operated by NOAA. A single satellite of the ITOS series exceeded the mission capabilities of both of the complementary ESSA satellites by furnishing complete global observations of Earth's cloud cover every 12 hours, while two TOS satellites required 24 hours to accomplish this [19]. Another improvement of the ITOS series was the capacity for taking pictures during nighttime by two infrared scanning radiometers. Additionally, the Solar Proton Monitor (SPM) was able to measure proton energy from the Sun, and a Flat Plate Radiometer (FPR) monitored Earth's heat balance. The primary and secondary data from the ITOS system for meteorological and scientific applications is listed in Table 4.1. Operational satellites of the ITOS series were subsequently renamed NOAA, launching a total of six satellites from 1970 to 1976.

**Table 4.1** Primary and secondary data from the ITOS satellite series

|                | Data                                      | Spectrum Range   | User                                                                  |
| -------------- | ----------------------------------------- | ---------------- | --------------------------------------------------------------------- |
| Primary data   | Daytime observations of cloud cover       | Visible spectra  | Directly transmit to users located around the world                   |
|                | Nighttime observations of cloud cover     | Infrared spectra | Directly transmit to users located around the world                   |
|                | Daily observations of global cloud cover  | Visible spectra  | For processing at the National Environmental Satellite Center of the ESSA |
|                | Daily observations of global cloud cover  | Infrared spectra | For processing at the National Environmental Satellite Center of the ESSA |
| Secondary data | Solar-proton density                      | N/A              | For processing at the National Environmental Satellite Center of the ESSA |
|                | Heat balance measurements                 | N/A              | For processing at the National Environmental Satellite Center of the ESSA |

In the early 1970s, the DMSP initiated a new series called Block 5A. These satellites introduced the Operational Linescan System (OLS) sensor, which provided images of clouds in both visual and infrared spectra. The OLS is an oscillating scan radiometer with low-light visible and thermal infrared (TIR) imaging capabilities, providing both day and night cloud cover imagery. Television resolution was also much improved, producing 2.7 km resolution data down-linked to ground sites, along with a small amount of 0.56 km high resolution data [20].

During 1970 to 1976, three Block 5A, five Block 5B, and three Block 5C satellites were launched on the Thor Burner II launch vehicle. Larger and more sophisticated Block 5D-1 satellites initiated with Block 5D-1/F01 launched in September 1976, and another four satellites were developed in the subsequent 4 years. In 1980, the fifth 5D-1 satellite was lost due to launch failure, and the operational 5D-1 satellites in orbit ceased to function prematurely. This series provided studies of environmental features, such as clouds, water bodies, snow, fire, and pollution in the visible/infrared spectra. According to the information recorded by scanning radiometers, observations of cloud height and type, land and water surface temperature, water currents, ocean surface features, ice, and snow could be achieved. Data were communicated to ground-based terminals, processed, interpreted by meteorologists, and ultimately used in planning and conducting U.S. military operations worldwide. More specifically, DMSP data were provided to the Air Force Global Weather Central (AFGWC), the Navy Fleet Numerical Oceanography Center (NFMOC) and to other civilian authorities through the DOC.

Since the DMSP and NOAA satellites were developed as completely separate systems for military and civilian use, respectively, there was a danger of partial duplication in funding two similar series. To avoid redundant development cost, NOAA was directed to adopt the DOD spacecraft design for the next generation of polar-orbiting satellites. In order to achieve daily round-the-clock global atmospheric sounding coverage, a two spacecraft NOAA system was used [1]. On October 13, 1978, the first polar satellite of the TIROS-N/NOAA series was successfully placed into orbit. Like the earlier TIROS systems, NASA took responsibility for the satellite only until proven operational.

This series was designed to provide higher resolution, day and night quantitative environmental data on local and global scales with technologically superior instrumentation than what was available on earlier ITOS/NOAA satellites. The orbits of this series were timed to allow complete global coverage twice per day per satellite in swaths of about 2,600 km in width. The TIROS-N imaging system, a four-spectral channel AVHRR, provided visible and infrared data for both daytime and nighttime. The AVHRR installed on later satellites offered capabilities for sea surface temperature determination, heat budget components estimation, and snow and sea ice identification. The satellite also carried an atmospheric sounding system called the TOVS-TIROS Operational Vertical Sounder, which provided vertical profiles of temperature and water vapor from Earth's surface to the top of the atmosphere. An SPM detected the arrival of energetic particles for use in solar storm prediction.

For military weather satellites, from August 1980 to December 1982, when the first Block 5D-2 satellite was successfully launched, meteorological data were supplied to the DOD entirely by civilian satellites. Nine Block 5D-2 satellites were launched from 1982 to 1997 on Atlas E and Titan II launch vehicles. The Space Systems Division began the procurement of six Block 5D-3 satellites from General Electric in 1989, which were later acquired by Lockheed Martin. In June 1990, the first 5D-3 spacecraft was scheduled to be delivered to the USAF [21].

The POES began with NOAA-K (NOAA-15), which was launched on May 13, 1998. With a suite of instruments, this series, like its precursors, was able to measure many parameters of Earth's surface and atmosphere. This included observations of cloud cover, incoming solar protons, positive ions, electron-flux density, and the energy spectrum at the satellite's altitude [22]. The satellites could also receive, process, and retransmit data from beacon transmitters and automatic Data Collection Platforms (DCPs) on land, ocean buoys, or aboard free-floating balloons. The primary instrument on POES was the 3rd generation instrument of AVHRR (AVHRR/3), which is the latest version of the sensor with six channels (three solar channels in the visible-near infrared region and three thermal infrared channels). NOAA-N Prime was the last in the series of Advanced TIROS-N (ATN) satellites launched on February 6, 2009. As previously mentioned, the National Polar-orbiting Operational Environmental Satellite System (NPOESS) was meant to be a replacement for both the DMSP and POES satellite programs.

However, on February 1, 2010, the Executive Office of the President dissolved and restructured the NPOESS partnership into separate development programs with two polar-orbiting satellite series to serve military and civilian uses. For civilian operations, the Joint Polar Satellite System (JPSS) would obtain environmental measurements from the afternoon orbit, while the Defense Weather Satellite System (DWSS) would have the morning orbit to satisfy Defense Department requirements. NOAA would lead and set requirements for the civilian and scientific community program that would provide global environmental data for Numerical

Weather Prediction (NWP) models for forecasts, as well as space weather observations. JPSS system data would be freely available to domestic and international
users, supporting U.S. commitments for the Global Earth Observing System of Systems (GEOSS). The DOD would take primary responsibility of the DWSS military
systems, which would provide continuity to the DMSP program.

The JPSS is currently planned to be a next-generation three-satellite program
for 2011 to 2029, which will provide operational continuity of satellite-based polar missions in the afternoon orbit that supports civil regional and global weather
and climate requirements. This will generate valuable oceanographic, environmental, and space weather information for operational users and scientists. Main payloads of this satellite will include five key instruments, a Visible/Infrared Imager/
Radiometer Suite (VIRRS), a Cross-track Infrared Sounder (CrIS), an Advanced
Technology Microwave Sounder (ATMS), an Ozone Mapping and Profiler Suite
(OMPS), and Clouds and the Earth's Radiant Energy System (CERES) [23]. The
DWSS satellite payload was planned to contain a VIIRS and Special Sensor Pre-
cipitating Electron and Ion Spectrometer (SEM-N) with an MIS for microwave im-
aging/sounding. However, the DWSS program was canceled in 2012 and the DMSP
follow-on is currently being redefined [24].

The NPOESS Preparatory Project (NPP) was launched on October 28, 2011,
with a Delta-II Mission Launch Vehicle from Vandenberg Air Force Base, Califor-
nia. Originally planned to be a proof-of-concept or demonstration satellite to test
key technologies, it was intended to act as a bridge between the EOS satellites and
the forthcoming series of JPSS satellites. It was subsequently renamed the Suomi
National Polar-orbiting Partnership (SNPP) in honor of Verner E. Suomi, a University of Wisconsin meteorologist, widely recognized as the Father of Satellite
Meteorology [25]. Serving as a risk-reduction and early-flight opportunity for the
JPSS program, it has a mission design lifetime of 5 years. The launch of the second
satellite, called JPSS-1, aboard a Delta II rocket is planned for early 2017 (Fig. 4.4).
Polar-orbiting data resulting from JPSS satellites will be entered into NWP models
for weather forecasts and climate monitoring. The JPSS represents a joint inter-
agency partnership between NASA and NOAA, with input from the DOD to ensure
an unbroken series of global weather data collection.

## Geostationary Orbiting Meteorological Satellites

The United States also maintains geosynchronous weather satellites that orbit the
equatorial plane of Earth at a speed matching Earth's rotation. This allows them to
hover continuously over one position on the surface at an altitude of about 35,800 km
above Earth. In the mid-1960s, the first successful experimental geosynchronous
communications satellites of the Syncom project and the first three operational meteorological satellites (ESSA-1 to ESSA-3) were successfully launched. On December 7, 1966, the first experimental equatorial synchronous satellite, called the

**Fig. 4.4 a** SNPP spacecraft and **b** JPSS-1 spacecraft of NOAA's next generation of polar-orbiting satellites. (Courtesy of NOAA) [26]

Applications Technology Satellite (ATS-1 launched on December 7, 1966), began its mission. The objectives of the ATS program were to test the experimental geostationary techniques of satellite orbit and motion, to measure the orbital environment at 35,800 km above Earth's surface, and to transmit meteorological information to surface ground stations [27].

In the ATS program, only ATS-1 and ATS-3 spacecraft successfully operated. Both carried a spin-scan camera, which provided high-quality cloud cover pictures. The camera system could capture a full-disk Earth every half-hour, affording a potential continuous watch of global weather and cloud patterns. ATS-1 was the first satellite to use frequency division access (FDMA) taking independently uplinked signals and converting for downlink on a single carrier. It carried out an impressive array of experiments in communications and collecting weather data.

ATS-3 was a larger spacecraft launched on November 5, 1967, and was the first satellite to routinely transmit full disk Earth-cloud imagery in color using a spin-scan cloud camera. By the early 1970s, ATS imagery was being used in operational forecast centers, with the first movie loops being used at the National Severe Storm Forecast Center (NSSFC) in 1972 [28]. Atmospheric motion depiction from the satellite was transferred into routine operations at the national forecast centers, and the resulting cloud motion vectors evolved into an important data source of meteorological information, especially over oceans [1].

The sixth and final ATS satellite, called ATS-6, was built by Fairchild Industries and launched on May 30, 1974. A much heavier and more sophisticated satellite than its predecessors, it was also the first educational satellite and first experimental Direct Broadcast Satellite that was also able to track sub synchronous S-Band satellites [29]. It was the first three-axis stabilized spacecraft in geostationary orbit and the first to use electric propulsion experimentally. During its five-year life, the ATS-6 retransmitted educational programming to India, the United States, and other countries, as well as conducting air traffic control tests and satellite-assisted search-

and-rescue missions. It carried an experimental radiometer that subsequently became a standard instrument aboard weather satellites.

In view of the success of the meteorological experiments carried aboard the ATS, NASA then fostered and developed the GOES system within NOAA that uses geosynchronous satellites for U.S. weather monitoring and forecasting. On May 17, 1974, the SMS-1 was the first geostationary meteorological satellite that was launched by NASA [30]. It was equipped with a Visible and Infrared Spin Scan Radiometer (VISSR) to provide high quality day/night cloud cover data and to take radiance-derived temperatures of the Earth/atmosphere system. In addition, a Space Environment Monitor (SEM) system was installed to measure proton, electron, solar X-ray fluxes, and magnetic fields, and weather facsimile data (WEFAX) could be relayed. A Data Collection System (DCS) similar to those on the NOAA polar orbiters was also installed. The SMS-2 was subsequently launched on February 6, 1975. The SMS-1 and −2, and GOES-1, -2, and -3 satellites were essentially identical.

In October 1975, NOAA's operation of a GOES series followed with the launch of GOES-1 satellite, which was spin stabilized and viewed Earth only about 10% of the time. It provided unique information about existing and emerging storm systems both day and night. GOES data soon became an indispensible part of NWS operations. The primary instrument on early GOES spacecraft continued to be the VISSR, which provided day and night observations of cloud and surface temperatures, cloud heights, and wind fields. More spectral bands were subsequently added to the VISSR, enabling the GOES system to acquire multispectral measurements from which atmospheric temperature and humidity sounding could be derived.

GOES-4 was launched in 1980 and enabled vertical temperature and water vapor profiling, permitting the monitoring of frame-to-frame movement of water vapor concentrations, leading to improved knowledge of global atmospheric circulation. GOES spacecraft operated as a two-satellite constellation about the equator, each observing about 60% of Earth's surface respectively and measuring atmospheric temperature and moisture, cloud cover, and the solar and geosynchronous space environment. Early GOES satellites were in operation from 1975 to 1994 [1].

In the spring of 1994, the first of the NOAA second generation operational geostationary satellites, GOES-I/GOES-8, was successfully placed into operation. It was the first three-axis stabilized spacecraft with the first fully independent sounder and a newly designed imager. The advanced spacecraft greatly enhanced the capability of the GOES system to continuously observe and measure meteorological phenomena in real-time, providing improved observational and measurement data around-the-clock. These enhanced operational services improved support for short-term operational weather forecasting and space environment monitoring, as well as atmospheric sciences research and development for numerical weather predication models, meteorological phenomena, and environmental sensor design.

To date, there have been 18 U.S. geostationary meteorological/environmental satellite launches: two SMS and 16 GOES satellites (with one launch failure). GOES 9 to 12 were successfully launched between 1994 to 2001. This generation of satellites viewed Earth 100% of the time and took continuous images and sound-

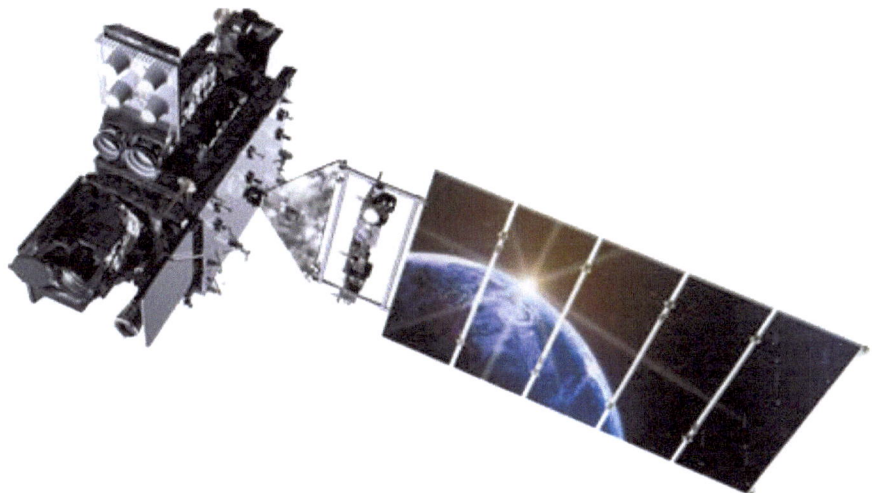

**Fig. 4.5** The GOES-R next-generation geostationary satellite. (Courtesy of National Science Foundation) [32]

ings. GOES-13 was launched on May 24, 2006, representing the first of the next generation of GOES satellites. The GOES N-P spacecraft had an advanced attitude control system using star trackers, a spacecraft optical bench, and improved Imager and Sounder instruments for navigation and registration [31].

For future geostationary satellites, GOES-R, is expected to be launched in 2015 with a mission design life until 2027. It will have a tremendous improvement on current GOES capabilities by including a suite of advanced instruments to provide higher spectral, spatial, and temporal information, real-time mapping of total lightning activity, improved severe weather forecasting, and solar monitoring capabilities. The GOES-R series of satellites will include GOES-R, S, T, and U, which is intended to extend the operation of the GOES satellite system through to 2036. GOES will continue to be collaboratively developed and acquired jointly between NOAA and NASA. A summary of the evolution of meteorological satellites in the United States is provided in Table 4.2, including all series and programs in chronological order.

The GOES-R series of satellites will extend the operation of the GOES satellite system over the next few decades. GOES satellites have been a critical component of this ongoing weather monitoring program, aiding forecasters in achieving more precise and timely forecasts. The United States operates two geostationary meteorological satellites, with one observing the East Coast and another observing the West Coast, thus achieving complete and overlapping national coverage. NOAA, with help from NASA, has established a continuous and long-term remote sensing capability for the nation based on a combination of polar and geostationary platforms.

This has proven useful in monitoring and predicting severe weather, such as tornadoes, tropical cyclones, and flash floods in the short term, and climate trends indicated by sea surface temperatures, biomass burning, and cloud cover in the

**Table 4.2** American meteorological satellites

| Satellite series | Launch period | Type | Missions |
|---|---|---|---|
| TIROS | 1960–1965 | Polar orbiting | NASA's first experimental step to determine if satellites could be useful in the study of Earth. Purpose to test various design issues for spacecraft: instruments, data, and operational parameters |
| TOS/ESSA | 1966–1969 | Polar orbiting | Initiated as extension of and a complement to the TIROS program. Primary objective to provide cloud-cover photography to the American NMC |
| ATS | 1966–1969 | Geostationary orbiting | Experimental communications satellites with the addition of other technology demonstrations, such as weather observation and investigation of the space environment |
| Nimbus | 1964–1978 | Polar orbiting | Objectives were to provide improved photographs of local cloud conditions by an APT system, and to evaluate an advanced vidicon camera system for daylight coverage and a high-resolution infrared radiometer system for nighttime cloud-cover observation |
| SMS | 1974–1975 | Geostationary orbiting | Operational prototypes of geostationary weather satellites to sense meteorological conditions from a fixed location |
| ITOS | 1970–1976 | Polar orbiting | Improved TIROS series continued mission of TIROS to provide continuous, day-to-day observations of Earth's weather systems |
| TIROS-N/NOAA | 1978–1994 | Polar orbiting | Aimed to provide higher resolution, day and night quantitative environmental data on local and global scales with technologically superior instrumentation |
| NOAA/POES | 1998–2009 | Polar orbiting | The POES mission is comprised of two polar orbiting satellites, which primarily provide long-range weather forecasting. Insure that non-visible data collection for any region of Earth is no more than 6 h old |
| GOES | 1975–2010 | Geostationary orbiting | Operational geostationary weather satellites. Aim is to monitor Earth's surface and space environmental conditions and to provide improved atmospheric and oceanic observations and data dissemination capabilities |
| DMSP | 1966–2009 | Polar orbiting | Mission is to provide global visible, infrared, microwave, and other secondary sensor data to military and civilian users. This data is used to produce real-time meteorological information and forecast products |

longer term. This system continues to improve with new technological innovations and sensors, resulting in improved observations of the weather system to protect property, health, public safety, and development.

# References

1. Davis, G.: History of the NOAA satellite program. J. Appl. Remote Sens. 1, 012504 (2007).
2. Rao, P.K.: Evolution of the weather satellite program in the U.S. Department of commerce—A brief outline. http://docs.lib.noaa.gov/rescue/TIROS/QC8795U47no101.pdf. (2001). Accessed 12 Aug 2013
3. National Research Council: Satellite Observations of the Earth's Environment: Accelerating the Transition of Research to Operations. National Academies Press, Washington, D.C. (2003)
4. NOAA: NOAA Historical Background. http://www.publicaffairs.noaa.gov/grounders/noaa-history.html. (2002). Accessed 12 Aug 2013
5. Whitney, P.L., Leshner, R.B.: The transition from research to operations in Earth observation: the case of NASA and NOAA in the U.S. Space Policy, vol. 20, pp. 207–215. (2004)
6. NOAASIS: NOAA's Geostationary and Polar-Orbiting Weather Satellites. http://noaasis.noaa.gov/NOAASIS/ml/genlsatl.html (2011). Accessed 12 Aug 2013
7. Air Weather Service: http://ngdc.noaa.gov/eog/pubs/DMSP_users_guide.pdf (1974). Accessed 12 Aug 2013
8. Los Angeles Air Force Base (LA AFB): Historical Overview of the Space and Missile Systems Center. http://www.losangeles.af.mil/shared/media/document/AFD-120802-071.pdf (2012). Accessed 12 Aug 2013
9. Hinnant, F.G., Swenson, H., Haas, J.M.: The NPOESS (National Polar-orbiting Operational Environmental Satellite System): a program overview and status update. IEEE. 1, 236–238 (2001)
10. NSIDC http://nsidc.org. Accessed 12 Aug 2013
11. The Space Review. The end of NPOESS. http://www.thespacereview.com/article/1601/1. Accessed 12 Aug 2013
12. Bristor, C.L., Ruzecki, M.A.: TIROS I photographs of the midwest storm of April 1, 1960. Month. Weather Rev. 316–326, vol. 88 (1960)
13. NOAA: http://noaasis.noaa.gov/NOAASIS/ml/40yearsa.html. Accessed 12 Aug 2013
14. Schnapf, A.: The TIROS decade. IEEE. 53–59, vol. 6 (1969)
15. NASA: http://science1.nasa.gov/missions/tiros/
16. NASA: http://nssdc.gsfc.nasa.gov/nmc/spacecraftDisplay.do?id = NIMBS–B. Accessed 12 Aug 2013
17. NASA: http://history.nasa.gov/SP-4312/ch5.html
18. National Reconnaissance Office: http://www.nro.gov/history/csnr/programs/docs/prog-hist-02.pdf. Accessed 12 Aug 2013
19. NASA:TIROS-M.http://ntrs.nasa.gov/archive/nasa/casi.ntrs.nasa.gov/19730006384_1973006384. pdf (1970). Accessed 12 Aug 2013
20. NSIDC: http://www.nsidc.org/dat/docs/daac/f13_platform.gd.html. Accessed 12 Aug 2013
21. Hall, R.C.: A history of the military polar orbiting meteorological satellite program, national recommanissance office. http://www.nro.gov/history/csnr/programs/docs/prog-hist-02.pdf (2001). Accessed 12 Aug 2013
22. NOAA Satellite Information System (NOAASIS): NOAA's geostationary and polar-orbiting weather satellites. http://noaasis.noaa.gov/NOAASIS/ml/genlsatl.html (2011). Accessed 12 Aug 2013

23. NOAA: http://www.jpss.noaa.gov. Accessed 12 Aug 2013
24. NASA: http://www.nasa.gov/mission_pages/NPP/main/index.html. Accessed 12 Aug 2013
25. WMO: http://www.wmo-sat.info/oscar/satelliteprogrammes/view/37. Accessed 12 Aug 2013
26. NOAA: http://www.jpss.noaa.gov/satellites.html. Accessed 12 Aug 2013
27. NASA Science Missions: http://science.nasa.gov/missions/ats/. Accessed 12 Aug 2013
28. Menzel, W.P., Purdom, J.F.W.: Introducing GOES-1: the first of a new generation of geostationary operational environmental satellites. B. Am. Meteorol. Soc. **75**(5), 757–781 (1994)
29. NASA: http://www.nasa.gov/centers/goddard/missions/ats.html. Accessed 12 Aug 2013
30. NASA: http://science.nasa.gov/missions/sms/. Accessed 12 Aug 2013
31. NOAA: http://www.oso.noaa.gov/history/operational.htm. Accessed 12 Aug 2013
32. National Science Foundation: http://www.sciencebuzz.org/topics/watching-above-way-above-career-spotlight. Accessed 12 Aug 2013

# Chapter 5
# European Meteorological Satellites and EUMETSAT

*All types of observation have a consistent and large positive impact on the 24-hour short-range forecast.*
—Dominique Marbouty, Director-General of the European Centre for Medium-Range Weather Forecasts (ECMWF) Council

This chapter provides an overview of European meteorological satellites and related ground systems, which help to deliver reliable and cost-efficient data, images, and products to the region. An historical overview is presented of satellites that were previously launched, as well as how meteorological satellite systems are currently established, maintained, and coordinated in Europe.

## First European Satellite Developments

After World War II, many European scientists left Western Europe to work in the United States or the Soviet Union. Although the majority of satellites were built by the two superpowers at the time, countries of Western Europe soon joined the race in space by becoming actively engaged in satellite development.

Ariel 1 (also known as UK-1 and S-55) launched on April 26, 1962 (Fig. 5.1). This satellite became the first British satellite and the first satellite in the Ariel program for studying the ionosphere. This made the United Kingdom the third country to operate a satellite, although the system was a joint project of NASA in the United States and the United Kingdom.

The San Marco 1 was the first Italian meteorological satellite and was built in-house by the Italian Space Research Commission. Under a cooperative plan, NASA provided rockets and launch crew training for Italian scientists and engineers. Launched on December 15, 1964, from Wallops Flight Facility, Virginia (USA), San Marco 1 became the first European-built weather satellite. This facility successfully conducted ionosphere or upper-atmosphere research. This became the first of five Italian-built weather satellites in the San Marco program. All of these satellites were launched using American Scout rockets. The last satellite called San Marco-D/L was launched on March 25, 1988. [2]

Astérix-1 (A-1) was the first French satellite, and it was launched on a French rocket on November 26, 1965, from Hammaguir, Algeria. This made France the

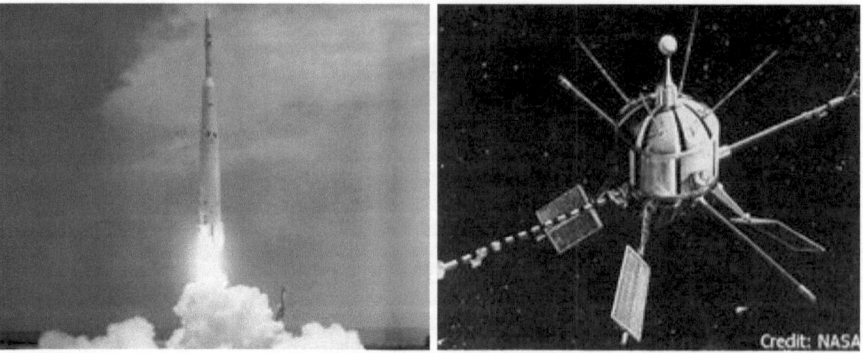

**Fig. 5.1** Launch of Ariel-1 on a Thor-Delta rocket and Ariel-1 in space. (Courtesy of NASA) [1]

third country to launch its own satellite, but the sixth to have its own satellite in orbit (after the United States, Russia, the U.K., Canada, and Italy). This satellite was thus launched and operated by the French government and specifically by CNES. CNES, which stands for Centre National d'Études Spatiales, was established in 1961. A-1 was intended to test the French Diamant rocket and to take measurements of the ionosphere. Unfortunately, the satellite only operated for two days. [3] A-1 is still in orbit and will be for centuries to come, since an antenna malfunction does not allow any commands to the satellite.

It soon became evident that although Western European countries could invest in research and space-related activities, solely national projects would not be able to compete with the major superpowers. European involvement in meteorological satellite programs became possible with the founding of the European Space Research Organization (ESRO) in 1962, which together with the European Launcher Development Organization (ELDO) became the precursor to the European Space Agency (ESA). ESA was founded in 1975 when the ELDO effort proved ineffective in developing a European launcher. It was thus decided that an integrated effort for research and development for Europe would likely be more successful, and thus the highly successful ESA was born.

Indeed the opportunities for European involvement in satellite meteorology were highlighted in a speech by David Arthur Davies, Secretary-General of the WMO, on September 11, 1967, to the Commission for Science and Technology of the Council of Europe in Geneva:

> I have always been surprised that the joint efforts of European countries in space activities have not been oriented towards a meteorological programme. It seems to me that such a programme contains all the appropriate elements to make it both attractive and fruitful. It would bring immediate practical advantages and would provide data of scientific interest. These advantages would accrue not only to the European countries, but also to many others. For example, if a geosynchronous meteorological satellite were to be placed in orbit above Africa, over the equator, it would provide highly significant information not only to Europe, but also on the whole African continent and would, in a certain sense, constitute a form of indirect technical assistance to the developing countries of this region.

> The cost of a programme of this kind would, I believe, be comparatively modest, if one considers the scale of other space enterprises. Last – and this would by no means be the least of the merits – the European countries would also be making a meaningful contribution to the achievement of the World Meteorological Organisation's plan, with the design and setting up of which they have been fully associated prior to approving it unanimously... [4]

In November 1968, delegates at the European Space Conference decided that ESRO would extend its brief to encompass satellite application programs with satellite meteorology as one of its prime foci. The ESRO took the initiative to create an Ad Hoc Meteorology Group (AHMG), which met for the first time in Geneva on May 1969. It conducted a feasibility study of a European operational meteorological satellite and concluded that an ideal system would include a geostationary and polar-orbiting satellite, as well as a communications package to collect and relay meteorological data from ground-based platforms. It was decided that the study would be devoted to taking measurements of: (a) the distribution of Earth's cloud cover both day and night, and (b) atmospheric temperature profile soundings. [5] It was further decided to concentrate on the Meteosat project, which was originally a French idea for a national geostationary satellite program.

## The Beginnings of Meteosat

Meteosat became the European meteorological program in geostationary orbit that was initiated in 1972 by the eight-nation ESRO. The initial Meteosat pre-operational program resulted in the launch of the first prototype Meteosat (Meteosat-1) by ESA on a Delta rocket from Cape Canaveral on November 23, 1977 (Fig. 5.2). Meteosat was also ESA's first Earth observation satellite, launched only two years after the official establishment of ESA. This was followed by Meteosat-2 launched in 1981 aboard an Ariane-1 rocket from Kourou in French Guiana, from where all remaining Meteosat missions were launched. The first generation includes Meteosat-1, -2, and -3, which were all developed and operated by ESA. Spacecraft design, instrumentation, and operation of this series were considerably similar to the American SMS/GOES satellites.

The overall objective of the Meteosat system was the provision of cost-effective satellite data and related services on a continuous basis to support the requirements of the member states. The data and services offered were mainly based on the requirements of operational meteorology, with the emphasis on supporting short-term weather forecasting, as well as supporting many other scientific research and applications, including fishing and agricultural production. Meteosat captured images every 30 min in three spectral channels, providing continuous and reliable meteorological observations from which a range of processed meteorological products were produced. [7] Later, second generation Meteosat missions were able to capture imagery every 15 min. This proved to be a significant milestone in Europe's era of satellite-based meteorology and set the stage for many subsequent successes for space-based monitoring for meteorology.

METEOSAT-1                                    FIRST IMAGE: 9 DEC 1977
                                                   COPYRIGHT ESA

**Fig. 5.2** Meteosat First Generation satellite prior to launch (left) and one of the first images captured by Meteosat-1 (right). (Courtesy of NASA) [6]

Meteosat was integrated into an international program sponsored by the WMO and the Global Atmospheric Research Program (GARP). GARP was developed in response to a U.N. resolution adopted in 1962 after the successes of the 1957/58 International Geophysical Year to develop meteorology services "for the benefit of all mankind." [8] Beginning in 1967 for 15 years that followed, and led by the WMO and the International Council of Scientific Unions, the primary goal of GARP field experiments was to help improve the quality of numerical weather prediction and to extend the forecast range up to 10–14 days.

GARP played a crucial role in pioneering influence in the use of satellites for continuous, global observation of Earth, leading a joint effort by the United States, the Soviet Union, Japan, and Europe to improve computer forecasting and modeling by using satellites in geostationary and polar orbits. Meteosat-1 was one of Europe's contributions to GARP and formed one element of the First GARP Global Experiment (FGGE). [9] GARP helped to form a unifying focus for the world's largely fragmented meteorological research programs, sealing commitment from participating states and marking the first global effort to gather data on atmospheric circulation and other weather phenomena.

## EUMETSAT

Throughout 1980, there were discussions about a suitable structure for managing operational satellite meteorology involving the heads of meteorological services across Europe. A key point in the evolution of European meteorological satellite systems was the recognition that it would be advantageous to create an organization specifically devoted to operational satellite meteorology. This organization would be responsible for developing programs to meet user requirements and negotiate funding from governments. The success of the Meteosat pre-operational program led a large group of European nations to establish firm plans for sustained operations by an independent inter-governmental agency, since ESA's mandate as a research and development organization did not allow for operational activities. [10]

In March 1983, an intergovernmental conference held in Paris agreed to form the European Organization for the Exploitation of Meteorological Satellites (EUMETSAT), which would assume financial responsibility for the Meteosat Operational Program (MOP) series of spacecraft, together with ground segment operations. On May 24, 1983, sixteen countries signed the EUMETSAT Convention, which came into force on June 19, 1986. The first EUMETSAT council meeting, hosted by ESA in Paris, agreed that the headquarters would be situated in Darmstadt, Germany, with John Morgan (U.K.) named as the first director. According to the preamble of the convention governing EUMETSAT, the main objectives of the organization are:

> To establish, maintain, and exploit European systems of operational meteorological satellites, taking into account as far as possible the recommendations of the World Meteorological Organization.
> A further objective of EUMETSAT is to contribute to the operational monitoring of the climate and the detection of global climatic changes. [11]

On January 1, 1987, responsibility for the operation of the Meteosat satellites was transferred from ESA to EUMETSAT, which formally took over direct responsibility for the reception, processing, dissemination, and archiving of European meteorological satellite data. It also became a member of the CGMS and took over responsibility for its secretariat. EUMETSAT became the owner of the Meteosat series beginning with Meteosat-4, which launched in July 1989 on Ariane Flight V29 (see Fig. 5.3), becoming the first geosynchronous satellite of the Meteosat Operational Program (MOP). At the same time, ESA continued to build and execute the MOP spacecraft series as a EUMETSAT contractor. Meteosat-3 became a spare satellite that was retained in orbit. This satellite eventually became the focus of controversial discussions when the United States requested it as a replacement for the NOAA failed GOES-6 geostationary satellite.

Meteosat-5 and −6 also came into operational usage. Meteosat-5 was launched on March 2, 1991. It operated over the Indian Ocean at 63°E longitude until 2007, supporting the Indian Ocean Experiment (INDOEX), which studied atmospheric pollution, solar radiation, and the interaction of clouds over the Indian Ocean region. The last spacecraft in the MOP series, Meteosat-6 was launched on November 20, 1993, with special operational service over Europe (retiring in 2006). This series provided rapid scan imagery with a repeat cycle of 10 min, but observed only a third of Earth's disk. Meteosat-6 also served as a back-up satellite for the service provided from 0° longitude.

The imaging system of Meteosat satellites was comprised of a radiometer working in the visible/infrared spectra. The generated images were distributed to the operational meteorological community in nearly real-time, where they were used for weather analysis and forecasting purposes. From the satellite imagery,

**Fig. 5.3** Meteosat-4 undergoing integration testing in 1989. (Courtesy of Colorado State University) [12]

quantitative products (e.g., cloud track winds and sea surface temperatures) were derived at the European Space Operations Center (ESOC), which served as the main mission control center for ESA. Although the first three MOP satellites supported all the nominal missions, all suffered from some anomalies with their imagery. Although these could be compensated for by on-ground software to a certain extent, switching between satellites several times each year was required to maintain optimum operations.

In May 1991, the decision was made that EUMETSAT would establish its own independent ground segment to replace the system established by ESA in 1977. In 1995, EUMETSAT implemented the Meteosat Transition Program (MTP), which was the first program fully developed under EUMETSAT control, including provision and launch of additional satellites, development of a new ground system, and handling routine operations. This covered the period from the phasing out of the MOP to the start of the second generation satellite program.

On November 15, 1995, EUMETSAT gained control of the Meteosat satellites in orbit, following 18 years of ESA operations. [13] Meteosat-7, the only satellite in the MTP program was launched on September 2, 1997, which had a similar design and capability to its predecessors. Meteosat-1 to Meteosat-7 satellites are thus often referred to as the first generation of Meteosat satellites.

The Meteosat Second Generation (MSG) system was established in June 1994 with a first launch in 2002 and operational services in 2004, under cooperation between EUMETSAT and ESA to ensure the continuity of meteorological observations from geostationary orbit. [14] Provisions were made for the procurement

**Fig. 5.4** MSG-1 satellite
in preparation for launch.
(Courtesy of ESA) [19]

of four geostationary meteorological satellites and provision for their launch and operation until 2020.

The first of the second generation missions began with the launch of MSG-1, later renamed to Meteosat-8, on August 28, 2002, from the Kourou launch site in French Guiana. The primary objective of MSG is to ensure continuity of atmospheric and space observation from geostationary orbit at 0° longitude and no equatorial inclination, as part of a worldwide, operational meteorological satellite system, consisting of four polar-orbiting and five geostationary satellites.

MSG satellites are designed for the needs of now-casting applications and numerical weather prediction. The spacecraft body is cylindrical-shaped, 3.2 m in diameter and 2.4 m high (see Fig. 5.4). In order to maintain its pointing toward Earth, its sensors are constantly spinning in an anti-clockwise motion at 100 RPM in relation to its platform that spins at the exact same speed in the clockwise motion in GEO orbit at an altitude of 35,870 km. [15] Primary payload instruments, include the Spinning Enhanced Visible and InfraRed Imager (SEVIRI) and the Geostationary Earth Radiation Budget (GERB) instruments, which support operational forecasting needs and climate studies, respectively. There is much greater capability when compared to the first generation satellites. The MSG provides twelve imaging channels instead of three, a higher spatial and temporal resolution, as well as extra services, such as improved measurement instrument and communication services. [16] They also provide more frequent and comprehensive data collection, with a baseline repeat cycle of 15 min. MSG satellites make up the EUMETSAT flagship program and provide improved data downloads that contain almost twenty times as much information as that from the first generation of satellites. [17] The MSG program includes a series of four identical satellites, Meteosat-8 to Meteosat-11 (MSG-1 to MSG-4).

As of 2013, Meteosat-7, -8, -9, and -10 were all operational, while Meteosat-11 (MSG-4) was planned for launch by early 2015. EUMETSAT in cooperation with ESA is preparing for a Meteosat Third Generation (MTG) system. [18] The MTG series is to be comprised of six satellites with the first spacecraft likely to be launched in 2020, along with the development of expanded ground segment and satellite operations. The system will include four enhanced imaging satellites. MTG-I will carry the Flexible Combined Imager (FCI) and an imaging lightning detection instrument called the Lightning Imager (LI). Two sounding satellite platforms (MTG-S) will carry an interferometer called the Infra-red Sounder (IRS), and the Ultraviolet Visible Near-infrared (UVN) sounder among other instruments. The platform will be three-axes body stabilized with instruments pointing at Earth for 100 % of their orbit time. Addition of a second sounding satellite platform is a key innovation of the new MTG program. This capability will enable not just imaging of weather systems but also the ability to profile the atmosphere layers and to perform chemical composition studies.

## European Polar-Orbiting Weather Satellites

After the growth of Europe's meteorological capacity and mastering of geostationary orbit, the need for a polar-orbiting system soon became apparent, in order to provide more detailed measurements of the atmosphere and improved observational coverage. Considering the high cost of space systems and growing need for complete and accurate atmospheric data collected at regular intervals for numerical weather prediction and climate monitoring, the NOAA entered into discussions with EUMETSAT to provide continuity of measurements from polar-orbiting satellites. In November 1998, building upon the POES program, an agreement called the Initial Joint Polar System (IJPS) was established. This is a joint program undertaken between EUMETSAT and NOAA to provide global-orbiting coverage. [20] This program provides for two series of independent but fully coordinated polar-orbiting satellites, exchange of instruments and global data, cooperation in algorithm development, and plans for real-time direct worldwide broadcasting.

The EUMETSAT Polar System (EPS) is the European contribution to the IJPS. With a series of European and U.S. Sun-synchronous polar-orbiting satellites,

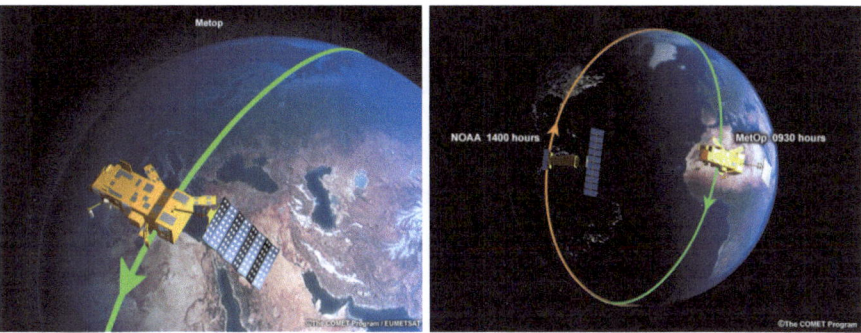

**Fig. 5.5** EUMETSAT EPS within the Initial Joint Polar System framework. (Courtesy of EUMET-SAT) [21]

Europe's EPS system takes over the morning orbit service from the current series of NOAA satellites, while NOAA's POES system (NOAA-18 and −19) continues to be responsible for the afternoon service (Fig. 5.5). In turn, the United States supplies legacy equipment, such as the Advanced Microwave Sounding Units (AMSU-A), AVHRR/3, HIRS, and SEM for incorporation into EPS satellite systems. The AMSU measures global atmospheric temperature, the HIRS measures global scene radiation in the infrared spectrum, the AVHRR captures global imagery in the visible/infrared spectra, and the SEM is applied for determining the intensity of Earth's radiation belts and the flux of charged particles at satellite altitude. With the implementation of the EPS, Europe contributes to the polar-orbiting component of the Global Observing System of the World Weather Watch.

The EPS program consists of the Meteorological Operational Satellite Program of Europe (MetOp) mission series of three polar-orbiting satellites. The spacecraft was primarily designed to provide continuous, long-term datasets, in support of operational meteorological forecasting and global climate monitoring. The MetOp series was also to provide enhanced monitoring capabilities to fulfill the requirements of studying Earth's climate system as expressed in a number of international cooperative programs, such as GCOS (Global Climate Observing System), IGBP (International Geosphere and Biosphere Program), and WCRP (World Climate Research Program). [22] It is complementary to the NOAA POES system, the EUMETSAT/ESA MSG system and the ESA Environment Satellite (ENVISAT) system.

The first MetOp satellite (MetOp-A) was successfully launched from Baikonur Cosmodrome, Kazakhstan, on a Soyuz launch vehicle on October 19, 2006, into a Sun-synchronous morning orbit. MetOp-B was launched on September 17, 2012, to replace MetOp-A. The MetOp series of spacecraft was jointly developed by EUMETSAT and ESA. At just over 4,000 kg in mass, it is Europe's second largest Earth observation satellite, after ENVISAT, which was launched in 2002. [23] Subsequent satellites in the series are planned for launch at approximately five-year intervals with a six-month overlap between consecutive satellites. The third satellite in the series, MetOp-C, is planned for launch in 2016/2017 from Guiana Space Center. [24]

**Table 5.1** European meteorological satellites

| Satellite Series | Launch Period | Type | Missions |
|---|---|---|---|
| Meteosat First Generation (MFG) | 1977–1988 | Geostationary orbiting | Primary mission is to provide high-resolution imagery of the Indian Ocean and surrounding areas day and night. |
| Meteosat Operational Program (MOP) | 1989–1997 | Geostationary orbiting | Improved design to provide important operational services with the retransmission of Data Collection Platform messages. It was also part of the Tsunami Warning System for the Indian Ocean area. |
| Meteosat Second Generation (MSG) | 2002–2012 | Geostationary orbiting | Primary mission is to ensure continuity of atmospheric observation from geostationary orbit at 0° longitude and inclination over Europe and Africa, as part of a worldwide operational meteorological satellite system. |
| European Meteorological Operational (MetOp) | 2006–2012 | Polar orbiting | Primary mission is to provide continuous, long-term data sets in support of operational meteorological and environmental forecasting and global climate monitoring. |

Since its launch, the MetOp series of satellites have demonstrated the importance of its atmospheric measurements delivered by an array of instruments. As previously mentioned, many sensors were 'heritage' instruments provided by the U.S. as its contribution to IJPS. However, new generation European developed instruments were also flown on MetOp satellites, including most notably the Infrared Atmospheric Sounding Interferometer (IASI), which greatly benefited numerical weather prediction for meteorologists and climatologists. MetOp-A carried the Global Ozone Monitoring Experiment-2 (GOME-2), which was a scanning spectrometer measuring atmospheric water vapor. GOME-2 is able to derive the track of a volcano plume, which was demonstrated during the eruption of the Icelandic volcano Grímsvötn in May 2011. [25] Other instruments specific to MetOp satellites include the Advanced Scatterometer (ASCAT) and the Global Navigation Satellite System Receiver for Atmospheric Sounding (GRAS).

EUMETSAT is planning a follow-on European Polar System Second Generation (EPS-SG) program, which will provide continuity of meteorological observations from polar-orbiting satellites for users in the 2020 time frame. [26] Similar to the MetOp, EPS-SG satellites will continue to be in a Sun-synchronous, low-Earth orbit (about 817 km in altitude), and provide full-globe observations with revisit times of 12–24 hours, depending on the instrument. Feasibility studies are ongoing, with consultation with users and application experts to define requirements for supporting operational meteorology and climate monitoring. Table 5.1 summarizes mission information of all European series and programs.

# References

1. NASA: http://skyquestt.com/2012/04/innovation-daily/. Accessed 12 Aug 2013
2. Harvey, B.: Europe's space programme: to Ariane and beyond. Springer-Proaxis (2003)
3. Encyclopedia Astronautica: http://www.astronautix.com/craft/asterix.htm. Accessed 12 Aug 2013
4. ESA: http://www.ares.esa.int/index.php/meteosat/introduction. Accessed 12 Aug 2013
5. NASA NSSDC: http://nssdc.gsfc.nasa.gov/nmc/spacecraftDisplay.do?id = 1977–108A. Accessed 12 Aug 2013
6. NASA: http://goes.gsfc.nasa.gov/text/geonews.html. Accessed 12 Aug 2013
7. eoPortal: http://www.eoportal.org/directory/pres_MeteosatFirstGenerationMFGSpacecraft.html. Accessed 12 Aug 2013
8. The National Academies: http://www.nas.edu/history/igy/. Accessed 12 Aug 2013
9. UNESCO—Intergovernmental oceanographic commission secretariat: http://unesdoc.unesco.org/images/0002/000249/024993eb.pdf. Accessed 12 Aug 2013
10. EUMETSAT: http://www.eumetsat.int/Home/Main/AboutEUMETSAT/WhoWeAre/EUMETSATHistory/index.html. Accessed 12 Aug 2013
11. EUMETSAT: Convention for the establishment of a European organization for the exploitation of meteorological satellites (Darmstadt-Eberstadt: EUMETSAT) (1983)
12. Colorado State University: http://rammb.cira.colostate.edu/dev/hillger/geo-wx.htm. Accessed 12 Aug 2013
13. EUMETSAT: http://www.eumetsat.int/Home/Main/Satellites/MeteosatFirstGeneration/MissionOverview/index.htm?l = en. Accessed 12 Aug 2013
14. EUMETSAT: http://www.eumetsat.int/Home/Main/Satellites/MeteosatSecondGeneration/MissionOverview/index.htm?l = en. Accessed 12 Aug 2013
15. EUMETSAT Meteosat Second Generation Instruments: http://www.eumetsat.int/Home/Main/Satellites/MeteosatSecondGeneration/Instruments/index.htm?l = en. Accessed 12 Aug 2013
16. Battrick, B.: Meteosat second generation: the satellite development. ESA Publications, Noordwijk, The Netherlands. (1999)
17. eoPortal: http://www.eoportal.org/directory/info_MeteosatSecondGenerationMSGSpacecraft.html. Accessed 12 Aug 2013
18. EUMETSAT Meteosat Third Generation: http://www.eumetsat.int/Home/Main/Satellites/MeteosatThirdGeneration/index.htm?l = en. Accessed 12 Aug 2013
19. ESA: http://www.esa.int/Our_Activities/Observing_the_Earth/Meteosat_Second_Generation/Follow_the_launch_of_MSG-1_from_ESA_and_Arianespace_establishments. Accessed 12 Aug 2013
20. Munro, R., Perez-Albiana, A., Callies, J., Corpaccioli, E., Elsinger, M., Lefebvre, A.: Expectations for GOME-2 on the METOP Satellites. ERS/Envisat Conference, Gothenburg (2000)
21. EUMETSAT: http://www.eumetsat.int/eps_webcast/eps/print.htm#s1p2. Accessed 12 Aug 2013
22. eoPortal—MetOp Program: http://www.eoportal.org/directory/pres_MetOpProgram.html. Accessed 12 Aug 2013
23. ESA: http://www.esa.int/esaLP/SEMAETN7BTE_LPmetp_0.html. Accessed 12 Aug 2013
24. EUMETSAT: http://www.eumetsat.int/Home/Main/News/CorporateNews/801146?l = en. Accessed 12 Aug 2013
25. DLR: http://www.dlr.de/caf/en/desktopdefault.aspx/tabid-7218/12014_read-30935. Accessed 12 Aug 2013
26. EUMETSAT: http://www.eumetsat.int/Home/Main/Satellites/EPS-SG/index.htm. Accessed 12 Aug 2013

# Chapter 6
# Russian, Chinese, Japanese, and Indian Meteorological Satellites

*An increasing number of nations operate meteorological satellites to meet the unprecedented demands of society for weather, water, and climate-related information.*
—Michel Jarraud, Secretary-General of the WMO

Apart from meteorological satellites operated by the United States and Europe, satellites have also been launched and operated by Russia, China, Japan, and India, contributing to global environmental and atmospheric observations. These weather satellites operate in low-Earth orbits, polar orbits, and geostationary orbits. On a collective basis, they provide invaluable information to meteorologists and climatologists for a range of applications, such as weather forecasts and natural disaster monitoring.

By international agreement, data sent from civil weather satellites are not encrypted and can be received and processed by anyone with the proper type of equipment. These systems are part of the Global Observing System (GOS) promoted by the WMO for global weather forecasting, which contributes to the GEOSS coordinated by the GEO. This chapter provides an overview of Russian, Chinese, Japanese, and Indian meteorological satellite systems, reviewing their strengths and capabilities.

## Russian Meteorological Satellites

The history of the Russian satellite program can be traced back to the world's first successfully launched artificial satellite Sputnik-1. During the Cold War, maintaining up-to-date information about weather conditions was necessary for supporting military strategic planning. On October 30, 1961, the Soviet government issued a decree ordering the development of an experimental meteorological network known as the Meteor satellite series, which became the nation's first operational polar-orbiting satellites for environmental and weather monitoring (Fig. 6.1) [1].

Today, within the Russian Federation, the sponsoring agency of the meteorological program is the Russian Federal Service for Hydrometeorology and Environmental Monitoring (ROSHYDROMET). This entity is the national central

**Fig. 6.1** The Sputnik-1
spacecraft was the first
artificial satellite successfully
placed into orbit. (Courtesy
of NASA NSSDC) [2]As

agency that provides relevant environmental and climate information to the public, various industrial organizations and decision-making bodies. It is a service in the Ministry of Natural Resources and Environment and it oversees public services related to meteorological and other geophysical processes. The Meteor series began in 1969 and was designed and developed by the All-Russian Scientific and Research Institute of Electro-mechanics (VNIIEM) of Moscow.

Prior to the Russian meteorological program, experimental scientific satellites were launched with the Kosmos series. The designation Kosmos is a generic name given to a large number of Soviet, and subsequently Russian, satellites, the first of which was launched in 1962. As of 2013, Russia has launched more than 2,485 Kosmos satellites with some Sputnik and Meteor satellites also given the Kosmos designation. The first experimental orbiting meteorological satellite was Kosmos-44 launched on August 28, 1964. It was the first in a series of prototype weather satellites used mainly to test basic spacecraft hardware, including a three-axis attitude control system, cremnium-based solar panels, and thermal protection systems. Kosmos-44 transmitted the first TV images of cloud cover.

A series of nine analogous Kosmos experimental satellites were launched from 1965 to 1969. These satellites served as test vehicles to develop and improve prototypes of TV and infrared (IR) cloud cameras and actinometric instruments [3]. The subsequent launches of Kosmos-122, Kosmos-144, and Kosmos-156 formed the first Soviet experimental weather-forecasting network. Interim experimental weather satellite instrumentation included two vidicon cameras for daytime monitoring, a high-resolution visible and infrared radiometer for day and night imaging, as well as an array of narrow- and wide-angle radiometers. These radiometers were used to measure the intensity of radiation reflected from the clouds and oceans, the surface temperatures of Earth and cloud tops, and the flux of thermal energy from the Earth-atmosphere system into space [4]. Assembly and testing of satellites continued to take place at VNIIEM, while analogous Kosmos satellites were launched until 1968, when the succeeding series was officially named Meteor-1 in 1969.

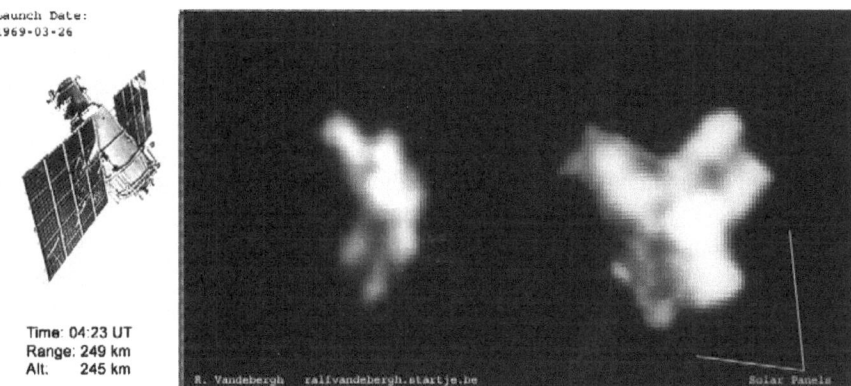

**Fig. 6.2** The Meteor 1-1 spacecraft (*left*). Image of the satellite (*right*) obtained from a range of 250 km captured on March 20, 2012, about 5 days before its predicted re-entry. The spacecraft body and solar panels are visible. The satellite fell in Antarctica on March 26, 2012, after more than 4 decades in orbit. (Courtesy of Space Safety Magazine) [7]

## Meteor Satellites

Originally developed in the 1960s, the Meteor satellite series was designed to monitor atmospheric and sea-surface temperatures, humidity, radiation, sea ice conditions, snow cover, and cloud patterns. The designations of the series Meteor-1, -2, and -3 define different payload configurations, improved spacecraft platforms, and different orbits [5].

In total, the Meteor-1 program launched 25 satellites into polar-orbit between 1969 and 1978 [6]. The satellites were designed to last for 6 months and frequently replaced. This resulted in an average of 3.4 launches per year for these satellites. On-board sensors consisted of the television-type optical instrument, TV infrared instrument, and a radiation budget sensor. Somewhat similar to the NOAA-TIROS series, the Meteor series provided a daily weather review for more than two-thirds of the globe, observation of cloud patterns, ice cover, and atmospheric radiation.

Meteor 1-1 was the first operational weather satellite of the Meteor series launched on March 3, 1969, with a main mission of cloud observation. This satellite had a mass of 1,300 kg with a length of 5 m and diameter of 2.5 m. The Meteor 1-1 spacecraft was placed in an orbital altitude of 560 km with two large solar panels attached to the sides (Fig. 6.2). The satellite carried two vidicon cameras for daytime photography, a scanning high-resolution infrared radiometer for day and night photography, and an actinometric instrument for measuring Earth's radiation field. Data were transmitted directly to ground receiving stations and analyzed by the Hydrometeorological Center in Moscow.

The successors of Meteor 1-1 include the Meteor-2, Meteor-Priroda, Meteor-3, and Meteor-3M. The second generation Meteor-2 series comprised of 21 operational meteorological satellites launched starting in July 1975 over a period of 8 years until 1993 with improved design and performance compared to Meteor-1 [8].

Operational design lifetime was planned to be extended from 6 months to 1 year with direct data transmission to receiving ground stations. Both Meteor-1 and -2 series were launched into non-Sun-synchronous polar orbit with an inclination of 81 to 82°.

The main improvement represented by the Meteor-2 was the introduction of infrared temperature/humidity sounding radiometric sensors that were able to build global vertical temperature maps. The instrument package contained three TV visual/infrared scanners, a five-channel scanning radiometer, and a Radiation Measurement Complex (RMC) device for measuring radiation flux densities in near-Earth space. The Meteor system provided fast-reaction weather forecasting and was able to distinguish characteristics of ice cover even in the Arctic Ocean, which proved to be useful for ship navigation and military planning.

At the beginning of the 1970s, gradual demilitarization of the Meteor program had begun with the goal of orienting space technologies towards civilian remote sensing purposes. The Meteor-Priroda series consisted of six satellites intended to demonstrate new instrumentation, mainly for multispectral land observation. The satellites were actively spin-stabilized over three axes with a design lifetime extended to 2 years of operation. The first satellite called Meteor-P1 was launched on July 9, 1974, while Meteor-P2 was launched from Plesetsk on June 29, 1977, becoming the first Soviet satellite to reach Sun-synchronous orbit.

The third generation Meteor-3 satellites were first launched in October 1985 and continued being launched during the 1990s with six operational meteorological satellites in total. Measuring 1.4 m by 4.2 m in dimension, the satellites had a three-axis stabilization attitude control system enabling orientation accuracy to be up to 0.5°. The satellites were equipped sophisticated instrumentation with multiple remote sensing instruments in addition to weather-forecasting payloads. Since the satellites were placed into a higher altitude than the previous series, an extension of the instrument swath width with complete coverage of Earth's surface was achieved [5]. The last two satellites of the series hosted foreign instruments for ozone and Earth radiation budget observations. For example, the fifth Meteor-3 satellite was launched on August 15, 1991, carrying the Total Ozone Mapping Spectrometer (TOMS), as the first and last U. S.-built instrument to fly on a Soviet spacecraft.

The follow-up Meteor-3M program only involved one Meteor-3M-1 satellite, which was launched on December 10, 2001, significantly delayed by financial problems. It featured sensor improvements, such as a 1.4 km resolution visible channel and a ten-channel radiometer with 3 km resolution. The satellite were also in Sun-synchronous orbit and included experiments, such as a Stratospheric Aerosol and Gas Experiment (SAGE III) payload designed to measure vertical profiles of aerosol, ozone, and other constituents of the atmosphere. It also carried a set of multi-spectral scanning sensors and other instruments designed to measure temperature and humidity profiles, clouds, surface properties, and high energy particles in the upper atmosphere.

The Meteor-M series represents a follow-on to the polar-orbiting meteorological mission to Meteor-3M. These satellites are designed as the next generation of Russian meteorological satellites. Meteor-M-1 was launched on September 17, 2009, with six instruments, including imagers, sounders, and a radar imaging system carried on-board [9]. It is capable of gathering multispectral imagery in the visible range, as well as radar imagery of Earth's surface and to perform surveillance of Earth surface features and atmospheric conditions. The Meteor-M-3 satellite will be equipped with a new-generation phased antenna radar for ocean monitoring [1]. Russian space officials have promised to orbit as many as four Meteor-M satellites by 2015 and to begin launching a new-generation Meteor-MP satellites series beginning in 2016. The primary mission of Meteor-M is similar to those for NOAA/NPOESS and EPS/MetOp series.

## GOMS-Electro Geostationary Satellites

In addition to polar-orbiting meteorological satellites, Russia develops and operates geostationary spacecraft to monitor Earth and the space environment. Geostationary Operational Meteorological Satellite-1 (GOMS-1), also referred to as Electro-1 or Elektro-1, represents Russia's first series of operational meteorological satellites in geostationary orbit, which has also led Russia to join the international geostationary weather monitoring group [10]. The program began in 1994 and was developed by the Russian Space Agency (Roscosmos) and operated by the Federal Service for Hydrometeorology and Environmental Monitoring (RosHydroMet). The spacecraft and instruments were developed by VNIIEM as the prime contractor.

Primary objectives of the GOMS-1 spacecraft were to acquire, in real-time, television images of Earth's surface and cloud cover in the visible and infrared regions of the spectrum, to measure the radiation levels and magnetic field of the space environment, and to measure vertical temperature profiles and cloud cover. Payload instruments included a radiation measurement system and scanning TV radiometer. The communications system obtained and retransmitted information via Russian and international Data Collection Platforms (DCPs), which were then exchanged among ground stations to the user community [11]. The GOMS-1 spacecraft was launched on October 31, 1994, but experienced initial problems with attitude control and never became fully operational. Limited operational capability was recovered in 1996, although visible imagery was not able to be broadcast due to technical issues with the sensors.

Electro-L is a new generation geostationary meteorological mission under development by RosHydroMet, Roscosmos, and the Scientific Research Center of Space Hydrometeorology "Planeta". Before Electro-L, Meteor-M-1 was the only operational weather satellite in orbit. Russia relied on meteorological data provided by American and European weather agencies [12]. The launch of Electro-L-1 on January 1, 2011, from Baikonur Cosmodrome marked a substantial contribution to Russian weather forecasts. These geostationary satellites provided a new wide variety of data for weather analysis and forecasting on both a global and regional scale. It was the first Russian weather satellite to operate successfully in geostationary orbit.

The primary objectives of Electro-L are to provide multispectral imagery of the global and regional atmosphere in both visible and infrared frequencies, and to

**Fig. 6.3** Artist's rendition
of the Elektro-L spacecraft.
(Courtesy of eoPortal) [10]

collect heliospheric, iono-spheric, and magneto-spheric data, as well as providing
data for climate change and ocean monitoring. Data collection services were provided from self-timed DCPs to the ground segment. The main instrumental payload
is an optical imaging radiometer, which provides image data in three visible and
near-infrared channels and in seven infrared channels [13]. Unlike NASA/NOAA
GOES satellites, Electro-L captures images in the infrared, as well as the visible
near infrared spectrum, providing valuable information about cloud movement and
vegetation cover [14].

The design life of Electro-L satellites is expected to be 10 years with a replacement policy based on launching satellites at roughly five-year intervals. The second
generation geostationary weather satellite in the series, Electro-L-2 is expected to
be launched shortly on a Zenit-3F rocket from Baikonur [15]. Future generations
of Russian geostationary weather satellites are also planned, including the third-
generation Electro-L-3 and Electro-M series, which will contribute to Russia's goal
to strengthen its weather satellite network and provide accurate weather analysis
and forecasting both for its territory and worldwide. Overall missions of Russian
weather satellite programs are summarized in Table 6.1.

## *Okean*

Although not specifically developed as a weather satellite, the Okean Earth observation satellite program does provide some cloud and climate-monitoring capabilities. The Okean program is a joint Ukranian/Russian remote sensing program for
ocean monitoring, including sea surface temperature, wind speed, sea color, ice
coverage, and cloud coverage, and precipitation [16]. The first prototype oceanography satellite (Okean-OE-1) was launched on September 28, 1983, into Sun-synchronous near-circular orbit. The subsequent Okean-01 and Okean-O satellite series
have contributed significantly to sea navigation, fishery, and coastal shelf usage

**Table 6.1.** Russian meteorological satellites

| Satellite series | Launch period | Type | Missions |
|---|---|---|---|
| GOMS/Elektro | 1994–2011 | Geostationary orbiting | Primary objective was to acquire, in real-time, television images of Earth's surface and cloud cover in the visible and infrared spectra and to obtain and retransmit information and data via Russia and international Data collection platforms |
| Kosmos series | 1964–1969 | Non-Sun-synchronous polar orbiting | Experimental role in LEO orbit. Primary objective was to provide round-the-clock surveillance and measure the intensity of radiation reflected from clouds and oceans, the surface temperature of Earth and cloud tops, and the total flux of thermal energy from the Earth-atmosphere system into space |
| Meteor 1 series | 1969–1977 | Polar orbiting | These three series had similar missions during their operating period, mainly to provide global information on the distribution of cloud, snow, ice cover, and surface radiation temperature once or twice a day |
| Meteor 2 series | 1975–1993 | Polar orbiting | |
| Meteor 3 series | 1985–2001 | Polar orbiting | |
| Meteor 3M series | 2001 | Polar orbiting | Primary objective was to monitor and measure meteorological phenomena, atmosphere, and space environment |
| Meteor M series | 2009 | Polar orbiting | Designed to obtain data for weather forecasts, to monitor Earth's ozone layer and radiation conditions in the upper atmosphere, as well as to provide information on ice floes for maritime shipping in polar regions |

applications. In particular, the Multi-spectral Opto-Mechanical Scanner (MSU-M) has provided cloud monitoring and sea surface temperature measurements. Launch of future Okean satellites will help continue improving the accuracy of weather, climate, and ocean forecasts.

# Chinese Meteorological Satellites

China's meteorological satellite program is called Feng-Yun (FY), meaning wind and cloud, which consists of both polar-orbiting and geostationary series. The China Meteorological Administration's (CMA) National Satellite Meteorological Center (NSMC) was founded in 1971 and is tasked with the responsibility of satellite operations and developing the ground segment. It is authorized to develop and operate the national satellite meteorological service for addressing weather and climate issues. The NSMC's main responsibilities include developing the Chinese meteorological satellite system, operating the satellite system, providing an information service to disseminate satellite data for climate prediction, forecasts, and warning, and implementing satellite engineering project contracts [17].

The CMA contracts the China Aerospace Science and Technology Corporation (CASC) to develop, design, and launch the Feng-Yun satellites, while the NSMC maintains overall system operations [18]. The funding body in China is the Ministry of Aerospace. FY-odd numbers (FY-1, FY-3, etc.) are applied to generations of polar-orbiting satellites, whereas FY-even numbers (FY-2, FY-4, etc.) refer to the geostationary series. The Feng-Yun satellites have played important roles in oceanography, agriculture, forestry, hydrology, aviation, navigation, environmental protection, and national defense, in order to fulfill the public service and information needs of Chinese society and national economy. In these ways, the satellites have provided support to the Chinese economy and have helped to mitigate many natural disasters.

The Chinese meteorological satellite program also enhances its ability to contribute to and collaborate with the international community in terms of a global environmental satellite system. The Feng-Yun satellites form a necessary part of the GOS sponsored by the WMO [19]. Another subordinate body under the CMA is the National Climate Center (NCC) [20], which was founded in 1995. The NCC is involved with climate monitoring and diagnosis, climate prediction and impact assessment, and climate change. The NCC currently functions as China's regional WMO climate center. It created the Beijing Climate Center (BCC) in March 2003 to support this role [21].

## *FengYun1 Polar-Orbiting Satellites*

The main objectives of the Feng-Yun program are to establish a comprehensive operational meteorological satellite system with the combination of polar and geostationary orbits, as well as a ground monitoring and data sharing system. China's polar-orbiting satellite program began with the first-generation FY-1 series that were built by the Shanghai Institute of Satellite Engineering. Both spacecraft were three-axis stabilized and powered by two solar arrays.

FY-1A was successfully launched into Sun-synchronous orbit on September 7, 1988, and FY-1B on September 3, 1990, by Long March 4 (CZ-4) boosters from Taiyuan, China. The primary payload contains a Multichannel Visible Infrared Scanning Radiometer (MVISR) for multi-purposes imagery and a SEM for space weather observations. This instrumentation enabled the satellites to acquire global visible and infrared cloud imagery for weather forecasting. Although the two experimental satellites suffered from some malfunctions, they nevertheless laid a solid foundation for the next generation of operational polar-orbiting satellites.

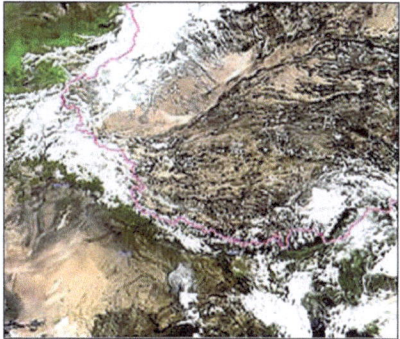

**Fig. 6.4** Illustration of the FengYun1 spacecraft (*left*) and the first image captured by FY-1D of northwest China (*right*). (Courtesy of eoPortal) [22]

With the experience from the first experimental satellites, the two operational satellites FY-1C and FY-1D represented upgraded versions. These satellites had improved imaging instruments, including the MVISR that now had ten channels. The improved design enabled more powerful observations of clouds, land, and ocean. The design life of these satellites was also extended. FY-1C was launched on October 5, 1999, and FY-1D was launched 2 years later on May 15, 2002 [22]. Examples of application areas of the observation data included the monitoring of drought, snow cover, floods, forest and grassland fires, dust storms, and sea ice.

## *FengYun2 Geostationary Satellites*

The Chinese geostationary meteorological program was initiated in the 1980s with the development of the FengYun2 (FY-2) satellite series by the NSMC and operated by the CMA. Its design involved dual spin-stabilized geostationary spacecraft with rotation velocity of 100 rpm [23]. The first experimental geostationary meteorological satellite FY-2A was successfully launched on June 10, 1997, with the follow-up experimental satellite FY-2B on June 26, 2000.

China's first operational geostationary weather satellite was FY-2C, which was launched on October 19, 2004, with a design life of 3–4 years [24]. The main satellite payload was the Stretched Visible and Infrared Spin-Scan Radiometer (S-VISSR), which is an opto-mechanical system with different image scanning modes. This satellite was capable of cloud imagery of five spectral channels (one visible and four infrared). This improved the satellite's capability for detecting and calculating water vapor contents, improved temperature resolution data, and improved ice cloud detection and sea temperature estimation. FY-2C was able to capture hourly visible, infrared, and water vapor disk images of Earth's surface. FY-2 spacecraft involve several subsystems for data circulation, including command and data acquisition, data transmission, weather facsimile, and data collection service.

**Fig. 6.5** Illustration of the
FY-2 geostationary satellite.
(Courtesy of eoPortal) [24]

Finally, the FY-2D spacecraft carries similar instruments as its predecessor. This satellite can generate daily image data and products used for environmental monitoring and weather forecasting services. Derived products include the atmospheric motion vector, sea surface temperature, and surface albedo [25]. Next generation FY-2 geostationary meteorological satellites are currently in their definition and pre-configuration stages with planned replacements until 2020.

## *FengYun3 Second Generation Polar-Orbiting Satellites*

FY-3 is the second generation and successor of the FY-1 series of Chinese Sun-synchronous polar-orbiting environmental satellites. This series is planned to include two experimental satellites and five operational satellites that will provide service until 2023. This series began with the launch of the experimental FY-3A satellite on May 27, 2008, into Sun-synchronous orbit. The series differs from FY-1 in that it involves atmospheric sounding, providing global air temperature, humidity profiles, and meteorological parameters. These satellites will thus measure cloud and surface radiation and increase the accuracy of weather forecasts. They can also be used for numerical weather forecasting [26]. These satellites will provide improved microwave imaging, higher spatial resolution imagery, higher temporal resolution, data acquisition, atmospheric composition detection, and radiation budget measuring capacity [27]. The FY-3 spacecraft have a design life of 3 years.

The mission of the FY-3 series is to obtain three-dimensional thermal and moisture profiles of the Earth's atmosphere, to monitor Earth's surface for meteorological forecasting and disaster warning, and to collect and retransmit data from DCPs. The experimental phase included two spacecraft launches with FY-3A on May 27, 2008 and FY-3B on November 4, 2010. The operational phase satellites are planned to be launched beyond 2013 and will have enhanced sounding and imaging capabilities with two FY-3 spacecraft in orbit, one in the morning time slot and one in the

**Table 6.2** Chinese meteorological satellites

| Satellite series | Launch period | Type | Missions |
|---|---|---|---|
| FY-1 | 1998–2002 | Polar orbiting | Main objectives were to: (a) acquire global surface and cloud imagery day and night; (b) measure surface and cloud-top temperature; (c) measure composition of orbit space environment |
| FY-2 | 1997–2008 | Geostationary orbiting | Primary mission was to gather visible, infrared, and water vapor cloud images and to carry out space environment monitoring |
| FY-3 | 2008–2012 | Polar orbiting | Main objectives were to: (a) obtain global measurements of 3-D temperature and moisture soundings of the atmosphere, and to measure cloud and precipitation parameters in support of numerical weather prediction; (b) collect global and local meteorological information |

afternoon. This will provide extra capability in monitoring large-scale meteorological disasters, natural hazards and environmental change, and monitoring climate variability. Planned replacements will occur at regular intervals until 2023.

## *FengYun4 Second Generation Geostationary Satellites*

FY-4 is the second generation geostationary meteorological satellite series currently in the definition and pre-configuration stages for 2016-2020. Compared to the FY-2 series, FY-4 satellites will include improvements in imaging instrumentation of 100-meter spatial resolution and atmospheric sounding capabilities [28]. This includes infrared hyper-spectral resolution atmospheric vertical sounding of temperature, humidity, and greenhouse gases. The series will also improve monitoring, "now-casting," and very short-range forecasting of meso-scale severe weather during the flooding season each year in China.

Chinese meteorological satellite programs are undergoing fast development to meet growing needs of national requirements and weather forecasting services. The Chinese meteorological satellite programs reflect the contribution made by China to the space-based global meteorological and environmental satellite system of the WMO, as well as the new GEOSS initiative. Meteorological satellite operators coordinate their activities on a global scale through participation in the CGMS, which meets once a year. The China Meteorological Administration has participated in the CGMS since 1989. After the United States and Russia, China has fast become the operator of the third largest national network of polar-orbiting and geostationary meteorological satellites in the world. Overall missions of Chinese weather satellite programs are summarized in Table 6.2.

## Japanese Meteorological Satellites

The Japan Meteorological Agency (JMA) is responsible for collecting and reporting weather data and forecasts in Japan, as well as observation and warning of natural hazards, such as earthquakes, tsunamis, and volcanic eruptions [29]. The JMA has been operating geostationary meteorological satellites since 1977, which has provided data for the prevention and mitigation of weather-related disasters, especially for monitoring typhoons and other weather conditions in the Asia-Oceania region.

### *GMS (Himawari) Series*

The Japanese Geostationary Meteorological Satellite (GMS) series, which is also known as "Himawari" (meaning a "sunflower"), was the first national satellite program of Japan for weather and environmental observations in geostationary orbit. It was operated by the JMA with the Japan Aerospace Exploration Agency (JAXA—formerly NASDA) as the spacecraft and launch service provider. The operational meteorological program consisted of five satellites in total, beginning with the launch of GMS-1 on July 14, 1977, from a NASA Delta rocket from Cape Canaveral. It was first launched as part of the GOS and forms an integral part of the WMO World Weather Watch (WWW) program. The GMS series has since provided valuable information for natural disaster prevention and meteorological services not only in Japan but also in the western Pacific and east Asia region [30].

There have been continuous efforts to maintain and enhance the GMS satellite series to provide continuous monitoring of significant meteorological phenomena in the region. The GMS spacecraft were spin-stabilized in geostationary orbit at a nominal position of 140°E longitude and with a design lifetime of 5 years. Mission objectives were to provide weather watch capabilities by a two-channel Visible-Infrared Spin Scan Radiometer (VISSR), collection of meteorological data from Data Collection Platforms (DCPs), direct broadcast of cloud images, and monitoring of solar particles with a SEM [31]. Development of the GMS series initially relied heavily on early U. S. GOES satellite design.

### *MTSAT Series*

With a design life of 5 years, GMS-5 was due to be replaced by a successor program called MTSAT (Multifunction Transport Satellite) in November 1999. The series was procured by the Japan Civil Aviation Bureau (JCAB) and JMA, while being funded by the Japanese Ministry of Land, Infrastructure and Transport (MLIT). However, due to launch failure of the H-2 vehicle on November 15, 1999, the MTSAT-1 satellite was destroyed and left Japan without weather satellite imagery.

**Fig. 6.6** Illustration of
MTSat-2 spacecraft in orbit.
(Courtesy of eoPortal) [33]

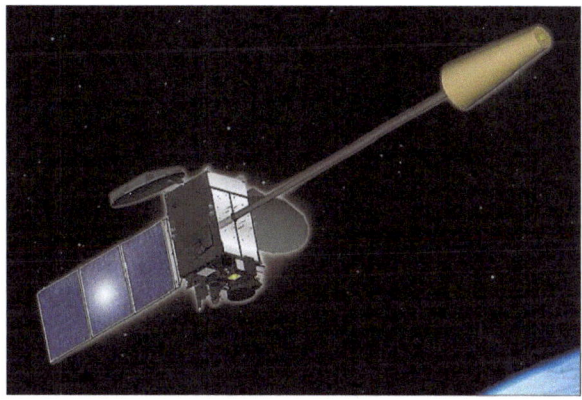

While the replacement satellite, MTSAT-1R, was being built, the U.S. agreed to lend Japan the GOES-9 geostationary environmental satellite as a back-up for GMS-5 (Hinawari-5) in order to fill the void and continue weather data collection from the western Pacific in case of satellite failure [31]. GOES-9 was moved over to the western Pacific and assumed the new GEO position at 155°E in April 2003. GMS-5 made its final observation on May 22, 2003, after which backup operations with the GOES-9 spacecraft began in cooperation between the JMA and NOAA/NESDIS.

The replacement spacecraft, MTSAT-1R (Himawari-6) was launched on February 26, 2005, on a H2A launch vehicle from Tanegashima Space Center, Japan with meteorological service operations switched over from GOES-9 to MTSAT-1R on June 8, 2005. MTSAT-1R performs full disk observations every 30 min with imaging channels consisting of a visible band and four infrared bands. Routine products of MTSAT, include an hourly atmospheric motion vector, clear sky radiance, cloud grid information, sea surface temperature, aerosol optical thickness, and snow/ice index (Fig. 6.6) [32].

MTSAT-2 (Himawari-7) was launched on February 18, 2006, into geostationary orbit and kept on standby mode until the end of the five-year MTSAT-1R mission (Fig. 6.7). On July 1, 2010, MTSAT-2 became operational for meteorological services, while MTSAT-1R is being kept on standby. MTSAT-2 has the same imaging channels as MTSAT-1R and provides similar imagery. The upcoming Hinawari-8 and Himawari −9 satellites are being prepared for launch in 2014 and 2016, respectively, to sustain and improve continuous satellite observations for the purposes of disaster prevention, weather forecasting, and now-casting. Data will help to improve the accuracy of numerical weather prediction and help to enhance climate and environmental monitoring. Newly developed products that are now planned include volcanic ash detection and height, and a global instability index. Information about Japanese meteorological satellites is summarized in Table 6.3, along with launch information and operation status.

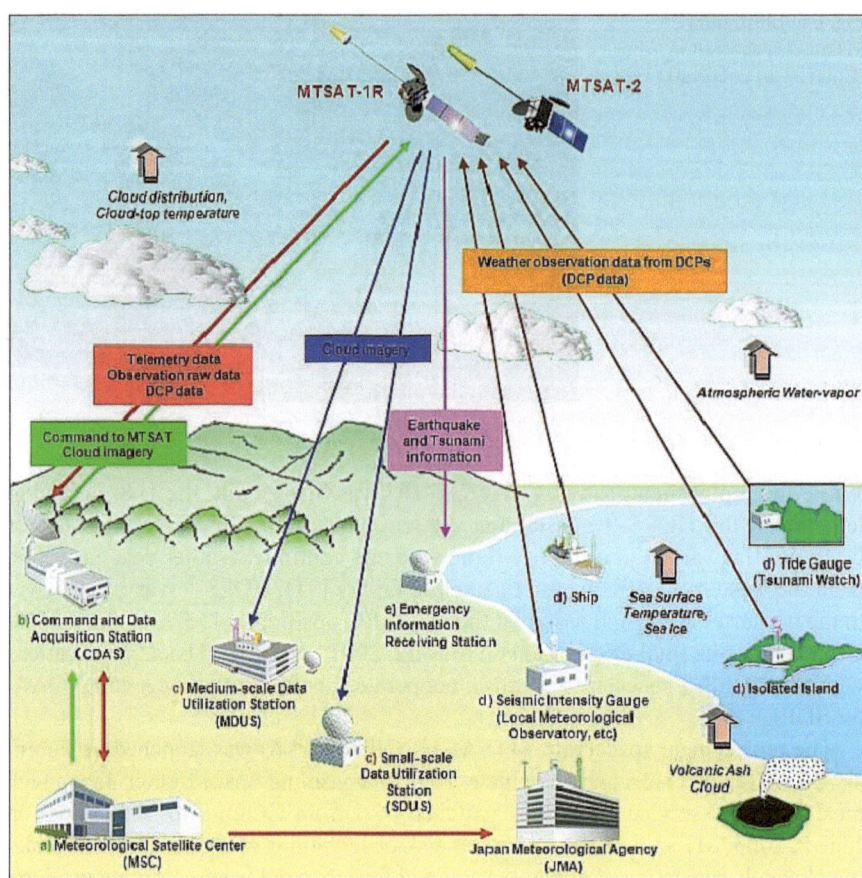

**Fig. 6.7** Overview of meteorological data communications for the MTSAT-1R and MTSAT-2 systems. (Courtesy of JMA) [34]

**Table 6.3** Japanese meteorological satellites

| Satellite series | Launch period | Type | Missions |
|---|---|---|---|
| GMS | 1977–1995 | Geostationary orbiting | Primary objective was to observe Earth's surface, cloud distribution, atmospheric water vapor distribution, solar particles, and other meteorological phenomena. Responsible for collecting meteorological data from data collection platforms and broadcasting cloud imagery |
| MTSat | 2005–2006 | Geostationary orbiting | Provided continuous operational services as GMS |

# Indian Meteorological Satellites

India has developed its own series of satellites for meteorological and environmental observation in both geostationary and polar orbits since the 1970s. The Indian Space Research Organization (ISRO) was established in 1969 as the primary space agency of the Indian government with the primary objective of advancing space technology and its applications for national benefit [35]. It is responsible for the development and operations of Indian satellites and instruments, while meteorological data is processed and disseminated by the India Meteorological Department (IMD) as the principal agency responsible for meteorological observations, weather forecasting, and seismology [36].

India's first satellite, Aryabhata, was launched by the Soviet Union on April 19, 1975, from Kapustin Yar using a Cosmos-3M launch vehicle [37]. The Rohini series of experimental satellites were launched by the ISRO from 1979 to 1983 to gain experience in building and operating satellites in space, as well as to test the Indian Satellite Launch Vehicle (SLV). These developments ushered in a revolution in India's space program to develop Earth observation satellites, such as the Bhaskara series for environmental and civilian applications, including telemetry, oceanography, and hydrology.

## *INSAT Series*

The INSAT (Indian National Satellite) system of the ISRO is the first multipurpose geostationary satellite system that was developed not only for meteorological applications but also for telecommunications, television broadcasting, and search and rescue services [38]. These satellites were initially manufactured by the U.S. Ford Aerospace and Communications Corporation (now known as Space Systems Loral) in accordance with Indian specifications. The overall objective of the mission was to perform round-the-clock surveillance of Earth and space environment, and to assist in predicting severe weather events in the Indian and Asia-Pacific region [39]. The program began as a joint venture between IMD, the Department of Space (DOS), the Department of Telecommunications (DOT), and All India Radio (AIR).

The INSAT series began with the launch of INSAT-1A in April 1982 on a Delta launch vehicle, and continued with three additional spacecraft launched within a period of 8 years (INSAT-1B on August 30, 1983, INSAT-1C on July 22, 1988, and INSAT-1D on June 12, 1990). The INSAT series was employed for meteorological observations over India and the Indian Ocean, as well as for telecommunication purposes. The on-board meteorology package consisted of a very high resolutions radiometer (VHRR) to provide full-frame imagery every 30 min. A DCS (Data Collection System) was also used to collect and relay environmental data (i.e., meteorological, hydrological, and oceanographic) to land- and ocean-based DCPs. Using the INSAT TV capability, early warnings of impending disasters, such as floods and storms, can directly reach the civilian population, even in remote areas.

**Fig. 6.8** Illustration of
INSAT-1A spacecraft.
(Courtesy of Colorado State
University) [40]

The second generation INSAT-2 series began in July 1992 with similar but improved payloads as INSAT-1. The series consisted of five spacecraft, INSAT-2A
to -2E, which have enhanced capabilities over the INSAT-1 series. This includes a
VHRR meteorological sensor that has a 2 km resolution VIS channel and 8 km resolution in the IR spectral channels. INSAT-2A was launched on July 9, 1992, with
replacements planned at roughly five-year intervals with two satellites always at
two nominal positions. INSAT-2E launched on April 2, 1999, on an Ariane 42P carrier rocket from Guiana Space Center, also carried a charge-coupled device (CCD)
camera capable of returning cloud images with a resolution of 1 km.

To date, meteorological services and telecommunication and television services
were frequently designed for combined payloads in the INSAT series. However,
advantages of developing separate satellites with payloads exclusively dedicated
to meteorology or telecommunication and broadcasting services were recognized.
Benefits included exploiting the scarce orbital slot and frequency spectrum, and
improving meteorological payloads in the absence of high power transponders [41].
There was also a need to develop an in-orbit replacement for meteorological services, since the VHRR/2 sensor on INSAT-2E had failed earlier in 1999.

MetSat-1 is India's first dedicated meteorological satellite launched by ISRO using a Polar satellite Launch Vehicle (PSLV) into geostationary orbit on September
12, 2002. It was later renamed to Kalpana-1 in honor of Dr. Kalpana Chawla—an
Indian-born NASA astronaut who had died in the space shuttle Columbia disaster.
Kalpana-1 was considered to be a cost-effective solution for providing meteorological payloads and services from geostationary orbit. This served to set the policy for
future INSAT missions to be configured separately for meteorological and communication payloads.

The next-generation INSAT-3 system began with the launch of INSAT-3B out
of six satellites developed in the series, with only three (INSAT-3A, D, D-prime)
equipped with meteorological instruments. The series includes the latest communication technology, such as high power antennae, steerable beam, and digital compression/decompression capabilities [41]. The latest INSAT-3D satellite is
configured exclusively as a meteorological satellite (Fig. 6.9). Its payload is complete

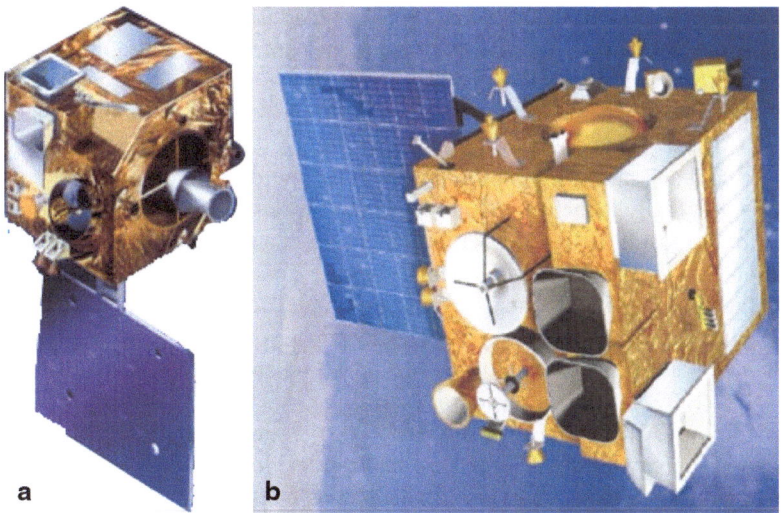

**Fig. 6.9** Illustrations of (**a**) India's first exclusive meteorological satellite, MetSat-1/Kalpana-1, and (**b**) INSAT 3-D spacecraft. (Courtesy of ISRO) [42]

with a meteorological imager, atmospheric sounder, search and rescue system, and data collection service. A summary of INSAT satellite payloads is listed in Table 6.4.

## *Indian Polar-Orbiting Satellites*

Complementing geostationary meteorological satellites, India has also developed a long series of sun synchronous polar-orbiting satellites beginning with the Indian Remote Sensing (IRS) system. This series started in the mid-1980s. The first generation satellites IRS-1A and -1B were launched successfully during 1988 and 1991 to provide Earth observation services for India. The purpose was to provide a continuous supply of synoptic, repetitive, multi-spectral data of the Earth's surface to support the national economy, agriculture, water resources, forestry, ecology, fisheries, and coastal management, similar to the U.S. Landsat program [43].

The initial program of Earth observation imaging was extended to other environmental applications with the addition of payload sensors and instruments. For example, a second-generation IRS-1C launched in 1995, and a similar IRS-P3 in 1996. These satellites carried imaging instruments to measure in four spectral channels, mainly in the visible and near-infrared, such as the Multispectral Opto-electronic Scanner (MOS).

IRS-P4 or OceanSat-1 was launched on May 26, 1999, and this satellite is dedicated exclusively to ocean monitoring. It was envisaged to provide service continuity for the operational users of Ocean Color Monitor (OCM) data, as well as assisting global weather forecasting. The OCM payload is a solid state camera designed to operate in eight narrow spectral bands. The camera is used to collect data on chlorophyll concentration, phytoplankton blooms, atmospheric aerosols, and particulate matter [44]. A second payload is a multi-channel scanning microwave radiometer (MSMR) for collecting data on sea surface temperature, wind speed, cloud water content, and water vapor content in the atmosphere above the ocean [45]. Although initially launched with a lifespan of 5 years, OceanSat-1 completed its mission on August 8, 2010, after serving for 11 years and 2 months.

OceanSat-2 was designed to provide follow-on services for Oceansat-1 with enhanced application potential for studying surface winds, ocean surface strata, chlorophyll concentration, phytoplankton blooms, atmospheric aerosols, and suspended sediments in water [46]. This satellite was launched on an Indian PSLV rocket on September 23, 2009, into a Sun-synchronous orbit. OceanSat-2 carries three payloads, including a OCM-2 similar to the device carried on Oceansat-1, a Ku-band pencil beam scatterometer (SCAT), and a piggyback payload called the Radio Occultation Sounder for Atmosphere (ROSA) developed by the Italian Space Agency. The SCAT is an active microwave device developed by ISRO to determine ocean surface level wind vectors by estimating radar backscatter, providing global ocean coverage and wind vector retrieval with a revisit time of 2 days [47]. The ROSA is an instrument that is designed to characterize the lower atmosphere and ionosphere [48].Using a Radio Occultation technique based on interactions of electromagnetic signals emitted between low-Earth Orbiting (LEO) satellites and a navigation satellite, such as Global Positioning System (GPS) satellites, ROSA is able to measure the vertical profile of atmospheric temperature, pressure and humidity [49].

The follow-up Oceansat-3 is currently in its preparation and development phase with planned launch in 2014. It will carry a thermal infrared sensor, twelve channel OCM, ocean scatterometer (OSCAT) for measuring sea surface wind vectors, and a passive microwave radiometer. The OCM and IR sensors will be used for studying potential fishing zones, as well as for ocean biology and sea state applications [50]. OceanSat satellites play an important role in forecasting the onset of the monsoon and its subsequent advancement over the Indian subcontinent and South-East Asia. It provides systematic data for oceanographic, coastal, and atmospheric studies.

In October 2011, ISRO and the Centre national d'études spatiales (CNES) began a cooperative experimental mission, called the Megha-Tropiques (Meteorological LEO Observations in the Intertropical Zone) to understand tropical meteorological and climatic processes, by obtaining reliable statistics on the water and the energy budget of the tropical atmosphere [51]. The French-Indian single-satellite experimental program successfully launched on October 12, 2011, into an LEO orbit over the equator. The purpose of the experiment is to study the convective systems (water cycle and energetic exchanges) that affect the Intertropical Convergence Zone (ITCZ), in particular in the latitudes between 10 and 20°. This satellite is expected to provide valuable data for climatic research [52].

Intertropical convergence zone measurements are achieved by a unique combination of scientific payloads on this French-Indian satellite. These four payloads include: (a) MADRAS (Microwave Analysis and Detection of Rain and Atmospheric Structures) instrument for measuring cloud properties and precipitation;

**Table 6.4** Indian meteorological satellites

| Satellite series | Launch period | Type | Missions |
|---|---|---|---|
| INSAT 1 series | 1982–1990 | Geostationary orbiting | Primary objective was to provide day-and-night surveillance service of weather events around the Indian region |
| INSAT 2 series | 1992–1999 | Geostationary orbiting | |
| INSAT 3 series | 2000–2003 | Geostationary orbiting | |
| KALPANA-1/ METSAT-1 | 2002 | Geostationary orbiting | Same as INSAT meteorological satellites |
| OceanSat | 1999–2002 | Polar orbiting | Primary objectives were to study surface winds and ocean surface strata, observe chlorophyll concentrations, and monitor phytoplankton blooms. Data obtained from this series are also used for weather forecast |
| Megha-Tropiques | 2011 | Polar orbiting | Primary objective was to study the convective systems affecting the Inter-tropical Convergence Zone, in particular in the latitudes between 10 and 20° |

(b) SAPHIR (Sounder for Atmospheric Profiling of Humidity in the Intertropics by Radiometry) instrument for humidity sounding; (c) ScaRaB (Scanner for Radiation Budget) instrument for measuring radiative fluxes; and (d) ROSA instrument similar to the device carried on OceanSat-2 [53].

Data collected from the Megha-Tropiques experiment are complementary to the data collected by other weather satellites in sun synchronous orbits, such as IRS satellites. The MADRAS microwave radiometer also contributes to the Global Precipitation Measurement mission joint with other agencies, such as JAXA and NASA. It further complements data from current regional monsoon projects, such as MAHASRI (Monsoon Asian Hydro-Atmosphere Scientific Research and Prediction Initiative) and the completed Global Energy and Water Cycle Experiment (GEWEX) Asian Monsoon Experiment (GAME) project [54, 55].The satellite lifespan is expected to continue past 2016 with substantial contributions made to atmospheric radiation research and space weather studies, and better understanding water cycle and energy exchanges in the tropics.

A summary of meteorological satellites series and programs developed by India are summarized in Table 6.4.

## Summary

Unprecedented demands of society for weather and climate-related information have led to nations developing their own geostationary and polar-orbiting meteorological satellite systems. Although the oldest and most developed meteorological satellite systems are those of the U.S. and Europe, other nations, including Russia,

China, Japan, and India, have invested in evolving their meteorological satellite-sensing capabilities. Significant advances have been attained in spacecraft design, sensor technologies, data exchange, and operations.

Furthermore, international coordination and cooperation is required to standardize format and display of vital meteorological observation data for distribution and sharing for weather forecasting and severe weather warnings. This is required to maximize the contribution of satellite observations for socio-economic development, reducing disasters, and addressing climate change. The deployment of new generations of meteorological satellite systems will provide a further boost to weather forecasting needs and accuracy.

# References

1.  Russian Space Web http://www.russianspaceweb.com. Accessed 12 Aug 2013
2.  NASA NSSDC http://nssdc.gsfc.nasa.gov/nmc/spacecraftDisplay.do?id=1957-001B. Accessed 12 Aug 2013
3.  NASA NSSDC http://nssdc.gsfc.nasa.gov/nmc/spacecraftDisplay.do?id=1964-053A. Accessed 12 Aug 2013
4.  NASA NSSDC http://nssdc.gsfc.nasa.gov/nmc/spacecraftDisplay.do?id=1967-039A. Accessed 12 Aug 2013
5.  eoPortal https://directory.eoportal.org/web/eoportal/satellite-missions/m/meteor. Accessed 12 Aug 2013
6.  WMO http://www.wmo-sat.info/oscar/satelliteprogrammes/view/97. Accessed 12 Aug 2013
7.  Space Safety Magazine http://www.spacesafetymagazine.com/2012/03/22/meteor-1-views-historic-spacecraft. Accessed 12 Aug 2013
8.  Hillger, D., Toth, G.: Un-manned satellites on postage stamps: the Meteor and FY-1 series. Astrophile. **48**(275), 177–181 (2003)
9.  Milekin, O.: Meteor-M and Elektro-L data access http://www.wmo.int/pages/prog/sat/meetings/documents/ET-SUP-6_Doc_09-02_ROSH.pdf (2011). Accessed 12 Aug 2013
10. eoPortal https://directory.eoportal.org/web/eoportal/satellite-missions/e/electro-1. Accessed 12 Aug 2013
11. OSCAR http://www.wmo-sat.info/oscar/satellites/view/72. Accessed 12 Aug 2013
12. Rianovosti http://en.rian.ru/science/20110121/162230923.html. Accessed 12 Aug 2013
13. Asmus, V.V., Dyaduchenko, V.N., Milekhin, O.E., Uspensky, A.B.: Proceedings from the 2005 Eumetsat Meteorological Satellite Conference. Remote sensing products and applications: Roshydromet Program. Croatia (2005)
14. E0Portal https://directory.eoportal.org/web/eoportal/satellite-missions/e/electro-l. Accessed 12 Aug 2013
15. OSCAR http://www.wmo-sat.info/oscar/satellites/view/74. Accessed 12 Aug 2013
16. eoPortal https://directory.eoportal.org/web/eoportal/satellite-missions/o/okean. Accessed 12 Aug 2013
17. National Satellite Meteorological Center http://www.nsmc.cma.gov.cn. Accessed 12 Aug 2013
18. Zhang, W., Xu, J., Dong, C., and Yang, J.: China's current and future meteorological satellite systems. Earth Sci. Satell. Remote. Sens. **1**, 392–413 (2006)
19. OSCAR http://www.wmo-sat.info/oscar/satelliteprogrammes/view/51. Accessed 12 Aug 2013
20. NCC http://ncc.cma.gov.cn/cn/. Accessed 12 Aug 2013
21. BCC http://bcc.cma.gov.cn. Accessed 12 Aug 2013

22. eoPortal  https://directory.eoportal.org/web/eoportal/satellite-missions/f/fy-1. Accessed 12 Aug 2013
23. WMO http://www.wmo-sat.info/oscar/satelliteprogrammes/view/52. Accessed 12 Aug 2013
24. eoPortal  https://directory.eoportal.org/web/eoportal/satellite-missions/f/fy-2. Accessed 12 Aug 2013
25. NSMC http://www.nsmc.cma.gov.cn/newsite/NSMC_EN/Channels/100090.html. Accessed 12 Aug 2013
26. eiPortal  https://directory.eoportal.org/web/eoportal/satellite-missions/f/fy-3. Accessed 12 Aug 2013
27. Dong, C., Yang, J., Zhang, W., Yang, Z., Lu, N., Shi, J., Zhang, P., Lu, Y., and Cai, B.:An overview of a new Chinese weather satellite FY-3A. Amer. Meteor. Soc. vol. 90(10), 1531–1544 (2009).
28. NSMC http://www.nsmc.cma.gov.cn/newsite/NSMC_EN/Channels/100090.html. Accessed 12 Aug 2013
29. JMA http://www.jma.go.jp/jma/jma-eng/satellite/index.html. Accessed 12 Aug 2013
30. JMA http://www.jma.go.jp/jma/en/Activities/brochure201003.pdf. Accessed 12 Aug 2013
31. eoPortal  https://directory.eoportal.org/web/eoportal/satellite-missions/g/gms. Accessed 12 Aug 2013
32. Tsunomura, S.:Current Status and Future Plan of Japanese Meteorological Satellite Program. http://jma-net.go.jp/sat/data/web/aomusc-2/Session1/1-3_STsunomura.pdf (2011). Accessed 12 Aug 2013
33. eoPortal   https://directory.eoportal.org/web/eoportal/satellite-missions/m/mtsat.  Accessed 12 Aug 2013
34. ISO http://www.isro.org. Accessed 12 Aug 2013
35. IMD http://www.imd.gov.in. Accessed 12 Aug 2013
36. ISRO http://www.isro.org/satellites/aryabhata.aspx. Accessed 12 Aug 2013
37. Goyal, S., 2011. INSAT-KALPANA data access. http://www.wmo.int/pages/prog/sat/meetings/documents/ET-SUP-6_Doc_09-06_INSAT-Kalpana.pdf. Accessed 12 Aug 2013
38. eoPortal   https://directory.eoportal.org/web/eoportal/satellite-missions/i/insat-2.  Accessed 12 Aug 2013
39. Colorado State University http://rammb.cira.colostate.edu/dev/hillger/geo-wx.htm. Accessed 12 Aug 2013
40. eoPortal https://directory.eoportal.org/web/eoportal/satellite-missions/k/kalpana-1. Accessed 12 Aug 2013
41. eoPortal   https://directory.eoportal.org/web/eoportal/satellite-missions/i/insat-3.  Accessed 12 Aug 2013
42. ISRO http://www.isro.org/satellites/satelliteshome.aspx. Accessed 12 Aug 2013
43. Joseph, G., and Deekshatulu, B.L.:Evaluation of remote sensing in India. In: Verma, R.K. (ed.) Space in pursuit of New Horizon.National Academy of Sciences publication, 331–354 (1992)
44. ISRO http://www.isro.org/satellites/irs-p4_oceansat.aspx. Accessed 12 Aug 2013
45. Narayanan, M.S., and Sarkar, A., 2001. Observation of marine atmosphere from Indian OceanSat-1. Proceedings from the 2001 Megha-Tropiques 2nd Scientific Workshop. Paris
46. http://isrohq.vssc.gov.in/isr0dem0v2/index.php/satellite/overview-satellites/40-satellites-details/227-oceansat-2. Accessed 12 Aug 2013
47. http://space.skyrocket.de/doc_sdat/oceansat-2.htm. Accessed 12 Aug 2013
48. eoPortal  https://directory.eoportal.org/web/eoportal/satellite-missions/o/oceansat-2. Accessed 12 Aug 2013
49. Agenzia Spaziale Italiana http://www.asi.it/Rosa/RosaEN/ROSA.htm. Accessed 12 Aug 2013
50. ISRO:Eleventh Five Year Plan Proposals 2007–12 for Indian Space Programme. http://planningcommission.nic.in/aboutus/committee/wrkgrp11/wg11_subspace.pdf (2006). Accessed 12 Aug 2013
51. CNES http://smsc.cnes.fr/MEGHAT. Accessed 12 Aug 2013

52. eoPortal https://directory.eoportal.org/web/eoportal/satellite-missions/m/megha-tropiques
53. Joshi, P.C.: Indian meteorological satellite missions: Current and planned. http://www.goes-r.
    gov/downloads/GOES_Users_Conference_V/GUC_V_slides/231645 %20GOES-R-presen-
    tation-PCJoshi.pdf (2008). Accessed 12 Aug 2013
54. http://mahasri.cr.chiba-u.ac.jp/index_e.html. Accessed 12 Aug 2013
55. http://www.hyarc.nagoya-u.ac.jp/game/. Accessed 12 Aug 2013

# Chapter 7
# International Collaboration in Meteorological Satellite Systems

*The atmospheric sciences require worldwide observations and, hence, international cooperation ... we shall propose further cooperative efforts between all nations in weather prediction and eventually in weather control. We shall propose, finally, a global system of satellite linking the whole world in telegraph, telephone, radio and television.*
—President John F. Kennedy (U. N. General Assembly, September 25, 1961)

Weather systems move across national boundaries and do not recognize political borders. Global cooperation in data collection processing and dissemination of meteorological data and products is needed in order to predict weather and to provide early warnings of severe weather-related phenomena. Such international cooperation can potentially save lives and minimize damage to property. Sharing of data products is key to improving forecasts and early warnings for nations, diverse economic sectors, and individuals. International meteorological cooperation is thus important for better risk management, as well as for managing the impacts of climate change. In order to predict weather, modern meteorology relies on near instantaneous exchange of weather information and observations across the entire globe.

## World Meteorological Organization (WMO)

The U.N.'s WMO is an intergovernmental organization and specialized agency for meteorology (weather and climate), operational hydrology, and related geophysical sciences [1]. It began as the International Meteorological Organization (IMO), which was an organization founded in 1873 with the purpose of exchanging weather information among countries internationally. The WMO was formally established in 1950 and became a specialized agency of the U.N. in 1951. Jointly with other organizations, it coordinates efforts to meet the needs for climate information, such as for climate monitoring, climate-change detection, seasonal to yearly predictions and assessments of the impacts of climate change [2]. WMO also promotes research that improves understanding of Earth's weather and climate systems and assists formulation of global and regional strategies and related action plans.

S.-Y. Tan, *Meteorological Satellite Systems,* SpringerBriefs in Space Development, DOI 10.1007/978-1-4614-9420-1_7, © The author 2014

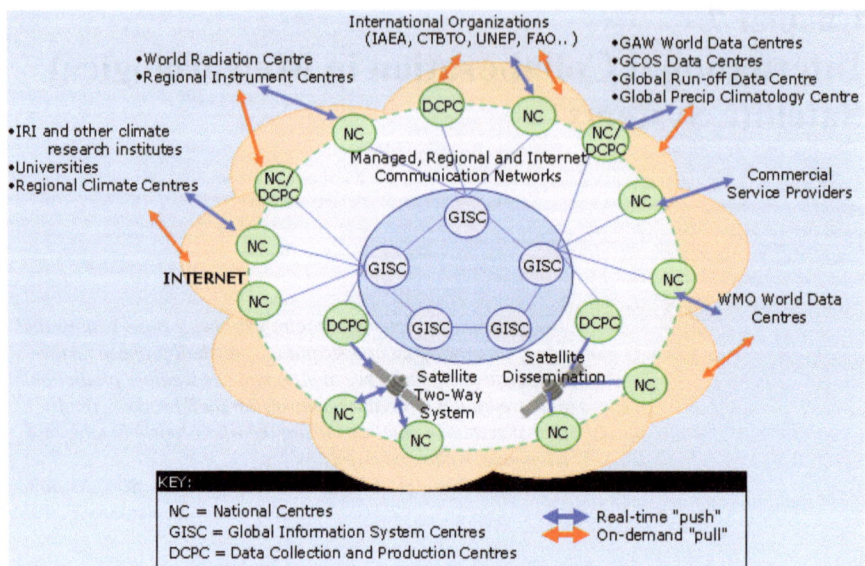

**Fig. 7.1** Diagram showing WIS centers in WMO member states for meeting information exchange needs. (Courtesy of WMO) [6]

The Convention of the World Meteorological Organization was initially signed on October 11, 1947, with the aim of coordinating, standardizing, and improving global meteorological activities. Sufficient ratifications were achieved to bring the organization into being as of 1950 [3]. As of January 2013, the WMO has 191 member states and territories. WMO provides a unique mechanism for the timely and unrestricted exchange of data, information, and products among its members that allows the production of real- or near-real time forecasts and early warnings. The universal availability of data, information, and derived products is facilitated via the WMO Information System (WIS), which ensures widely available and freely exchanged data between WMO centers and national weather services [4].

The WIS provides an integrated system for observing, data processing, data communication, and data management to provide public weather services, in response to requirements of WMO programs [5]. It incorporates three types of centers for providing information services, collecting and disseminating data and products between WMO members. As shown in Fig. 7.1, these are, (a) Global Information System Centers (GISCs) for regional and global dissemination, (b) Data Collection or Production Centers (DCPCs) responsible for collecting and generating data products and processed value-added information, and (c) National Centers (NCs) for data distribution on a national basis. The WIS enables time- and operational-critical data distribution and is implemented through telecommunication means (i.e., leased links, data networks, and satellite-based telecommunications) for timely delivery of forecast and warning services and products.

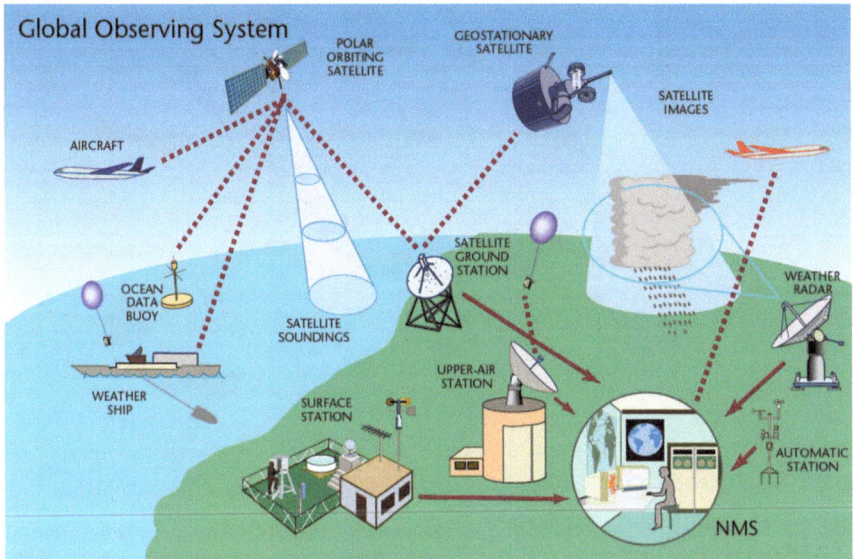

**Fig. 7.2** Components of the GOS that supports WMO programs. (Courtesy of WMO) [13]

The WMO carries out its work through a range of scientific and technical programs that assist Members in providing a range of meteorological and hydrological services. The WMO space program promotes and coordinates the availability
and use of satellite-derived observations and data products for weather, climate,
water, and related applications to WMO members [7]. It provides guidance on remote-sensing techniques for meteorology, hydrology, and climatology disciplines.
The WMO space program has four main components: (a) a space-based observing
system, (b) access to satellite data and products, (c) awareness and training, and
(d) space weather coordination.

The space-based observing system is made up of the space-based GOS [8], which
is also part of the WMO's WWW program that was first drafted in 1963. The GOS
provides atmospheric and ocean surface observations that are necessary for weather
analyses, forecasts, advisories and warnings, for climate monitoring and environmental activities. GOS is a coordinated system of Earth- and space-based observing facilities used for making meteorological and other environmental observations
and operated by national meteorological services within national or international
satellite agencies. It promotes wide availability and utilization of satellite data and
products to support weather, climate, water, and related environmental applications.
As shown in Fig. 7.2, GOS consists of various observation platforms and components to provide and exchange near real-time global information around the clock,
including observations from the ground, upper atmosphere, marine, aircraft-based,
weather radar, and space satellites. Data are collected by more than 11,000 land stations, 1,000 upper-air stations, 3,000 aircraft, and more than 1,000 ships working

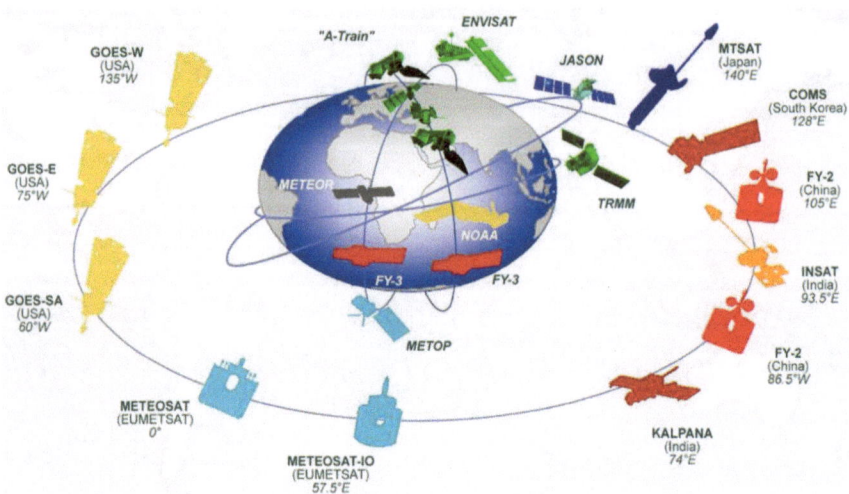

**Fig. 7.3** Schematic overview of the WWW's space-based Global Observing System. (Courtesy of WMO) [14]

in tandem with about 188 National Meteorological Centers and 50 Regional Specialized Meteorological Centers [9]. These are bolstered by about 16 operational meteorological and 50 environmental research satellites.

The core of the WWW program consists of the GOS, along with the Global Telecommunication System (GTS) and Global Data-Processing and Forecasting System (GDPFS). These collectively provide observing systems, telecommunication facilities, and data processing and forecasting centers to facilitate meteorological and environmental information exchange across the entire globe. The WWW has a key role in coordinating the collection and distribution of meteorological data for weather forecasts. Within the WWW, the WMO Integrated Global Observing System (WIGOS) applies the WIS to connect together all regions for data exchange, management, and processing. It focuses on providing an integrated, comprehensive, and coordinated framework to enable existing space-based observing systems to support data delivery and services for WMO members [10].

The WWW's GOS program is comprised of three types of satellites, including operational meteorological polar orbiting, geostationary, and environmental research and development (R&D) satellites. Figure 7.3 provides a schematic overview of the space-based GOS. Polar-orbiting and geostationary satellites are usually equipped with visible and infrared imaging and sounding instruments, which derive meteorological parameters. They can provide atmospheric profiles of temperature and humidity, and track cloud patterns and water vapor. R&D satellites comprise the newest constellation in the space-based component of the GOS. These satellites provide valuable data not normally observed by operational meteorological satellites, and can also provide new knowledge that can lead to technical improvements in operational systems.

The formation of GOS was historically based on the recognition that meteorological observations and weather services are strengthened by international

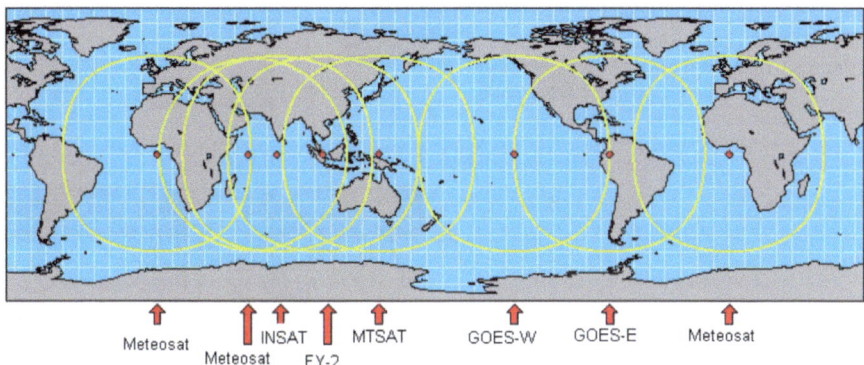

Meteosat          INSAT          MTSAT          GOES-W          GOES-E          Meteosat
          Meteosat          FY-2

**Fig. 7.4** Worldwide geostationary satellite coverage. (Courtesy of U. S. Department of Commerce, NOAA, and NESDIS, 2009) [15]

collaboration, coordination, and data sharing. In the early stages of meteorological satellite development, the early success of the TIROS program marked a major achievement for the global community of meteorologists. Data collected by TIROS was not just restricted to the United States but was available on a worldwide scale, including to Russia and China. With plans drafted for future geostationary satellites, the demand for international coordination of meteorological observations grew. These needs precipitated the formation of the WWW program.

Polar-orbiting and geostationary satellites for meteorological applications have advanced over the decades. The U.S. (GOES), Europe (METEOSAT), China (FY-2), India (INSAT), Japan (MTSAT), and Russia have all placed their own geostationary weather satellites into orbit, as shown in Fig. 7.4. The WMO has facilitated global planning for operational geostationary satellites and low-Earth orbit satellites, including developing a future vision for GOS to 2025 [11] and a Global Contingency Plan [12]. Apart from satellite observing systems, GOS also consists of *in situ* observing systems, telecommunication center and systems, and data processing and forecasting centers that are operated by WMO member states. The WWW acts as the backbone of worldwide operational observation and processing of meteorological data.

## Coordination Group for Meteorological Satellites (CGMS)

The CGMS was formed on September 17, 1972, and has held annual meetings since then [16]. It came into being when representatives from Europe, Japan, U.S., and observers from the WMO gathered in Washington, D.C., to discuss common interests relating to the design, operation, and use of geostationary meteorological satellites. CGMS provides an international forum for the exchange of technical information on geostationary and polar-orbiting meteorological satellite systems. It has expanded both in terms of its membership and objectives.

**Fig. 7.5** GSICS in the WMO
Global Observing System.
(Courtesy of WMO) [18]

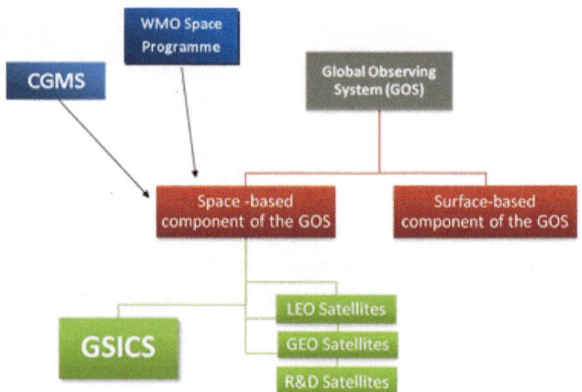

Membership currently consists of 15 member organizations and 6 observers. EU-METSAT has held the CGMS Secretariat since joining the group in 1987 and is responsible for organizing the annual CGMS plenary meeting with the support of the local CGMS member host.

According to the CGMS charter, the main objectives of the group are threefold [17]. First, the aim is to have a clear focus on coordination of long-term and sustainable satellite systems relevant to weather and climate to which both operational and R&D agencies contribute. Second, the CGMS provides a technical focus to the discussions among its members. Third, the CGMS seeks to respond as far as possible to requirements from WMO and related programs (e.g. WIGOS, GCOS) through close interaction with WMO. CGMS meetings have facilitated a better coordinated and more coherent approach for the operation of meteorological satellites.

CGMS has developed and coordinated some key initiatives, including the Global Space-based Inter-Calibration System (GSICS), the Sustained Coordinated Processing of Environmental Satellite Data for Climate Monitoring (SCOPE-CM), and the Virtual Training Laboratory (VLab). The GSICS is an international collaborative effort that was initiated in 2005 by WMO and CGMS to develop a coordinated approach to calibration and inter-calibration of operational weather satellites of the GOS (see Fig. 7.5; [18]). It aims to deliver calibration corrections required for accurate space-based observations and accurate integration of data from multiple observing systems into products, applications, and services. Calibration strategies are delivered by monitoring instrument performances, operational inter-calibration of satellite instruments, comparing measurements to absolute references and standards, and recalibration of archived data. The GSICS contributes to the integration of satellite data within the WIGOS and the GEOSS.

The SCOPE-CM was established in November 2008. Its mission is to develop a coordinated framework for generating climate data records from space observations [19]. This is especially important for operators of long-term satellite missions and for providing high quality long-term data sets for monitoring the ECVs defined by the GCOS [20]. This initiative is highly relevant for observing climate variability and modeling climate trends. Particular attention is given to fundamental technical aspects of sensors and satellites, such as instrument design, satellite orbits, and reliable retrieval algorithms for converting measured data into geophysical parameters. The SCOPE-CM is also establishing the sustained Climate Data Record (CDR) generation infrastructures and activities. The ultimate goal is to combine and harmonize data from different geostationary sensors in order to generate a long-term, homogeneous climate data record with global coverage for climate monitoring [21].

The VLab was established by the WMO and CGMS to improve education and training in satellite meteorology [22]. It is a global network of about 13 training centers and 8 meteorological satellite operators that work together to improve the use of data and products from meteorological and environmental satellites throughout WMO Member countries. The VLab aims to improve delivery of training services by providing supporting resources on meteorological satellite systems, data, products, and applications, and facilitating and fostering research and the development of socioeconomic applications at the local level though effective training programs.

## CEOS, GCOS, and GEO

International collaborative efforts also include the Committee on Earth Observation Satellites (CEOS), which was established in 1984 to coordinate and manage international civil space-borne missions designed to observe and study Earth [23]. CEOS is made up of its members, associates, and the Secretariat. It holds annual plenary meetings. Currently, 53 members and associate members participate in CEOS planning and activities, comprised mostly of space agencies, national, and international organizations. Associates are governmental organizations that are international or national in nature and currently have a civil space-segment activity. The permanent Secretariat is maintained jointly by ESA, EUMETSAT, NASA, NOAA, JAXA, and the Ministry of Education, Culture, Sports, Science and Technology of Japan (MEXT). Moreover, a Strategic Implementation Team (SIT) was formed to define, characterize, and develop the vision for CEOS participation in the Integrated Global Observing Strategy (IGOS) partnership created in 1998. The IGOS partnership was subsequently dissolved in 2008, but paved the way for the future formation of the GEO. CEOS remains recognized as the major international forum for the coordination of Earth observation satellite programs and for the interaction of these programs with users of satellite data worldwide [24].

The GCOS was established in 1992 and is an internationally coordinated system of observing systems for meeting the totality of national and international needs for climate and climate-related data and information [25]. It consists of the climate-relevant components of all established environmental observing networks and systems. It is co-sponsored by the WMO, the Intergovernmental Oceanographic Commission (IOC) of the United Nations Educational Scientific and Cultural Organization (UNESCO), the United Nations Environment Program (UNEP), and the International Council for Science (ICSU). The primary goal of the GCOS is to provide continuous, worldwide, reliable, comprehensive data and information on the state and behavior of the global climate system. It includes both *in situ* and remote-sensing components, with its space-based components coordinated by CEOS and CGMS.

From its inception, the GCOS has addressed core tasks, including defining and regularly updating climate requirements for global observing systems and highlighting gaps in global observing systems for climate monitoring. Specifically, the GCOS provides observations of the ECVs, which are needed to make significant progress in the generation of global climate products and derived information. The GCOS also supports all the components of the World Climate Program, the assessment role of the Intergovernmental Panel on Climate Change (IPCC), and the international policy development role of the United Nations Framework Convention on Climate Change (UNFCCC). The GCOS Climate Monitoring Principles (GCMPs) provide basic guidance regarding the planning, operation, and management of observing networks and systems, including satellites to ensure that high-quality climate data are available and contribute to effective climate information. The GCOS role in observing ECVs and monitoring climate change is discussed in further detail in Chap. 9.

The GEO is an intergovernmental partnership leading coordinated efforts to build a comprehensive, coordinated, and sustained Earth observation system, called GEOSS [26]. GEO was launched in response to calls for action by the 2002 World Summit on Sustainable Development and affirmed by the Group of Eight (G8) 2003 Summit for the urgent need for coordinated Earth observations to support decision making and management of natural resources. It is a voluntary partnership of governments and international organizations, providing a framework for developing new projects and coordinating their strategies and investments. Currently, the membership includes 88 governments and 67 intergovernmental, international, and regional organizations that have "Participating Organization" status. The GEO is constructing GEOSS on the basis of a ten-year implementation plan for the period of 2005 to 2015, which defines the vision and expected benefits of GEOSS [27].

The goal of GEOSS is to build upon existing national, regional, and international systems to provide comprehensive coordinated Earth observations from thousands of instruments worldwide. The collected data from GEOSS will be transformed into vital information to focus on nine "societal benefit areas" that are of critical importance to people and society, which include [26]:

- reducing loss of life and property from natural and human-induced disasters.
- understanding environmental factors affecting human health and well-being.
- improving management of energy resources.
- understanding, assessing, predicting, mitigating, and adapting to climate variability and change.
- improving water resource management through better understanding of the water cycle.
- improving weather information, forecasting, and early warning of severe weather.
- improving the management and protection of terrestrial, coastal and marine ecosystems.

- supporting sustainable agriculture and combating desertification.
- understanding, monitoring, and conserving biodiversity.

GEOSS is intended to provide decision-support tools to a wide variety of users, linking together existing and planned observing systems worldwide.

Since the inception of GEO, coordination of and investment in Earth-observation systems has increased, and there has been an accelerated trend towards full and open data sharing and improved data and information access. The GEO and GEOSS infrastructure will encourage more integration and dissemination of data sets produced by diverse systems and instruments. There is a growing trend to provide unrestricted access, often cost-free, to remotely sensed data. These data combined with information products and services are already empowering decision makers and managers, while also aiding policy leaders as they confront the complex and interlinked global challenges of the twenty-first century.

# References

1. WMO: http://www.wmo.int/pages/about/index_en.html. Accessed 12 Aug 2013
2. WMO: The World Meteorological Organization at a Glance. (2009)
3. WMO: Convention of the World Meteorological Organization, adopted by the Washington Conference. (1947)
4. WMO: http://www.wmo.int/pages/themes/wis/. Accessed 12 Aug 2013
5. WMO: Guide to WMO Information System (WIS). WMO-No. 1061. (2012)
6. WMO: http://www.wmo.int/pages/prog/www/WIS/centres_en.html. Accessed 12 Aug 2013
7. WMO: http://www.wmo.int/pages/prog/sat/index_en.php. Accessed 12 Aug 2013
8. WMO: http://www.wmo.int/pages/prog/www/OSY/GOS.html. Accessed 12 Aug 2013
9. WMO: http://www.wmo.int/pages/prog/www/OSY/Gos-components.html. Accessed 12 Aug 2013
10. WMO: http://www.wmo.int/pages/prog/www/wigos/index_en.html. Accessed 12 Aug 2013
11. WMO: Vision for the GOS in 2025. Commission for Basic Systems Fourteenth Session, Recommendation 6.1/1 (CBS-XIV). Dubrovnik, 25 March–2 April 2009
12. WMO: CGMS Global Contingency Plan, Version 2. (2007)
13. WMO: http://www.wmo.int/pages/prog/www/OSY/images/GOS-fullsize.jpg. Accessed 12 Aug 2013
14. WMO: http://www.wmo.int/pages/prog/sat/globalplanning_en.php. Accessed 12 Aug 2013
15. U.S. Department of Commerce, NOAA, and NESDIS: User's Guide for Building and Operating Environmental Satellite Receiving Stations. http://www.noaasis.noaa.gov/NOAASIS/pubs/Users_Guide-Building_Receive_Stations_March_2009.pdf. (2009). Accessed 12 Aug 2013
16. CGMS: http://www.cgms-info.org. Accessed 12 Aug 2013
17. CGMS: http://www.cgms-info.org/docs/general-publications/cgms-charter.pdf. Accessed 12 Aug 2013
18. GSICS: http://gsics.wmo.int. Accessed 12 Aug 2013
19. CGMS: http://www.cgms-info.org/initiatives/scope-cm. Accessed 12 Aug 2013
20. GCOS: http://www.wmo.int/pages/prog/gcos/index.php?name= EssentialClimateVariables. Accessed 12 Aug 2013
21. CGMS: http://www.cgms-info.org/initiatives/scope-cm/scope-cm-continuation. Accessed 12 Aug 2013
22. WMO: http://www.wmo-sat.info/vlab. Accessed 12 Aug 2013
23. CEOS: http://www.ceos.org. Accessed 12 Aug 2013

24. AthenaGlobal: Committee on Earth Observation Satellites (CEOS). (2004)
25. WMO: http://www.wmo.int/pages/prog/gcos. Accessed 12 Aug 2013
26. GEO: http://www.earthobservations.org. Accessed 12 Aug 2013
27. GEO: The Global Earth Observation System of Systems (GEOSS) 10-Year Implementation Plan. (2005)

# Chapter 8
# Evolving and Future Capabilities

> *Climate is what we expect,*
> *weather is what we get.*
>
> —Mark Twain (1835–1910)

To date, direct benefits derived from operational meteorological systems have been witnessed in meteorology, climatology, and oceanography applications. The rapid evolution of meteorological satellites led to improved observation of Earth's environment and more accurate forecasts of the weather, especially of hazards and extreme weather events. Changes in weather and climate can significantly impact our economy, infrastructure, and daily lives.

Weather phenomena, such as thunderstorms or fog, can pose a severe threat to safety or business efficiency. Timely forecasts and warnings are required to minimize the consequences. Now-casting, which involves mapping current weather in real-time and rapidly issuing weather warnings and developing solutions to support decision-makers, is one of the most challenging tasks for weather forecasting. In order for now-casting to be successful, very frequent and high-quality observation imagery of Earth and the atmosphere are required.

According to the WMO, now-casting seeks to provide detailed description of current weather conditions with forecasts obtained by extrapolation for a period of 0–6 hours ahead [1]. Based on radar, satellite, and other observational data, a forecaster is able to analyze small-scale weather conditions and forecast weather phenomena in a small area (i.e., a city) with high accuracy and high temporal resolution up to a few hours in advance. Now-casting can be used as a powerful location-specific forecasting tool for public warnings of severe weather or hazards, including tornados, thunderstorms, and tropical cyclones (see Fig. 8.1), which can lead to flash floods, lightning strikes, and destructive winds. This can contribute to reducing fatalities and injuries due to weather hazards, property damage, and improved safety and efficiency for industry, transportation, and agriculture.

Geostationary satellites are uniquely placed to deliver valuable information for numerical weather predication models that complement observations made by polar-orbiting satellites. They can observe winds and the displacement of clouds, as well as water vapor patterns, thus significantly enhancing climate monitoring capabilities. For example, the second generation European Meteosat satellites (MSG) exploit a two-satellite system with one satellite providing full disk imagery of the European and African continents and parts of the Atlantic and Indian oceans every

S.-Y. Tan, *Meteorological Satellite Systems,* SpringerBriefs in Space Development, DOI 10.1007/978-1-4614-9420-1_8, © The author 2014

**Fig. 8.1** Satellite data received and processed at the Earth Scan Lab, Louisiana State University, during Hurricane Lili in the Gulf of Mexico in 2002. Visible channel image from NOAA-16 with the storm track shown as a solid line. Data from NDBC buoys 42001 and 42003 were also employed. (Courtesy of NOAA) [3]

15 min, while the second satellite delivers more frequent images every five minutes over Europe only. This Rapid Scanning Service [2] combined with polar-orbiting observations provide further support for local thunderstorm warnings.

During the next decade, substantial changes are expected to be made to the NOAA constellation of geostationary and polar-orbiting satellites to improve performance, quality, and timeliness of weather data collection. The next-generation GOES-R satellite series will usher in a new era for the U.S. geostationary environmental remote-sensing program. The first satellite of this new series is expected to launch in 2015, and its ten-year period of dedicated service is expected to significantly improve the quality and timeliness of weather and other environmental observations. The new Advanced Baseline Imagery (ABI) will be carried aboard this satellite series, which is a 16-channel imager that will provide faster, more

**Fig. 8.2** The Geostationary Lightning Mapper (GLM) of the GOES-R satellite showing the assembly (left) and engineering design (right). (Courtesy of GOES-R) [8]

detailed scans and better products for forecasting, severe weather warning, numerical weather prediction, and climate and environment surveillance. It will provide cloud and moisture imagery for full disk, continental United States, and meso-scale coverage for monitoring, forecasting, and severe weather warning [4]. Compared to the current GOES imager (GOES-8 to -12), the improved ABI will contain four times higher spatial resolution, five times faster imaging, increased spectral coverage, and more accurate measurements for observing subtle features.

Another new development of the GOES satellite series is the Geostationary Lightning Mapper (GLM) [5], which will provide continuous lightning measures over most of the area covered by GOES-East and -West, including most of the tropical cyclone regions of the Atlantic and eastern north Pacific basins (see Figs. 8.2 and 8.3) [6]. GLM is an optical transient detector and imager operating in the near-infrared that maps total lightning (in-cloud and cloud-to-ground) flash rates and trends with near spatial resolution of 8–14 km continuously day and night. GLM data will improve local forecasts and warnings of severe weather, tornado warning and air quality, as well as providing storm intensification information [7]. Applications of GLM include airline and military routing, disaster preparedness, forest fire planning, climate studies, and severe thunderstorm forecasts and warnings, including providing advanced observations of potentially destructive thunderstorms over land and ocean areas.

In addition, GOES-R will further improve reports, alerts, warnings, and forecasts of potentially dangerous events, such as solar flares, geomagnetic storms, and other solar disturbances. Solar-pointed instruments such as the Solar Ultraviolet Imager (SUVI) and Extreme Ultraviolet Sensor/X-Ray Sensor Irradiance Sensors (EXIS) will aid in monitoring the highly variable solar and near-Earth space environment to protect life and property of those sensitive to space weather fluctuations. SUVI will produce images of the full solar disk every minute to detect abnormal activity and help provide early warning of solar events, thus replacing the current GOES-M/P

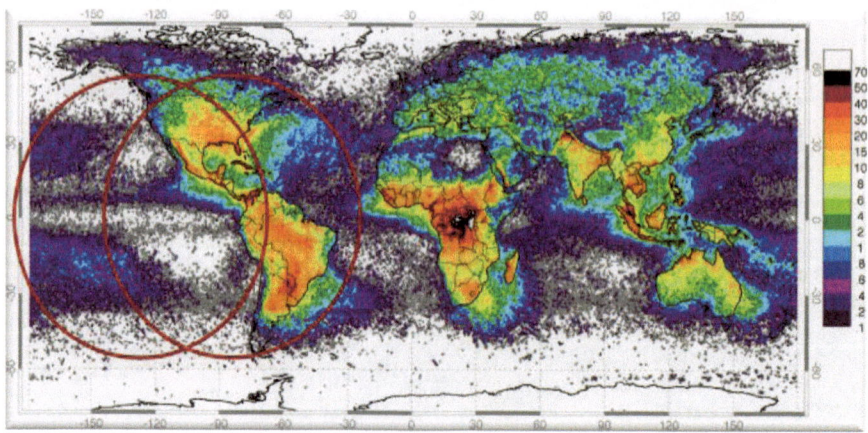

**Fig. 8.3** Global distribution of proxy lightning data from a combined nine years of observations of the NASA Overshooting Top Detection (OTD) (4/95-3/00) and LI (1/98-12/04) instruments. The GOES West and GOES East field-of-view are shown by *red circles*. (Courtesy of GOES-R) [9]

series Solar X-ray Imager (SXI) instrument. EXIS will continuously monitor solar output, measuring the full disk solar X-ray flux and monitoring the duration and magnitude of X-ray flares. Collectively, space weather instrumentation planned for GOES-R will continue the long history of observations from the GOES program, with expanded services and capabilities for solar and near-Earth space weather prediction.

The Joint Polar Satellite System (JPSS) is the next generation of U.S. polar-orbiting, non-geosynchronous environmental satellites, providing global environmental data for weather forecasts, oceanography, climate adaptation and mitigation, and coastal management through 2025 [10]. JPSS will replace the current NOAA POES satellite series and ground systems, exceeding previous capabilities in terms of quality, volume, accuracy, and timeliness of weather data products and services. A collaborative agreement between NOAA and EUMETSAT will have Europe's Polar System providing the mid-morning orbit, while JPSS covers the afternoon orbit. An emphasis has also been placed on real-time delivery of JPSS data products, providing data to the Direct Readout community through High Rate Data and Low Rate Data broadcasts. Direct broadcast of weather data will support real-time regional and time-critical applications. The JPSS-1 mission is currently scheduled for launch in December 2016, and the JPSS-2 mission in November 2021.

The U.S. and European meteorological satellite systems have had a close relationship for many years. In Europe, EUMETSAT works closely with ESA and other partners on the development of next-generation satellites, maintaining a continuity of services in both geostationary and polar orbits. After a consultation process with users and experts from member states and considering user requirements in the 2015–2035 timeframe, EUMETSAT and ESA have decided that the third generation Meteosat (MTG) system will remain focused on strengthening now-casting and very short-range forecast capabilities [11]. Major enhancements will include more frequent imagery acquisition (every 10 min for full disk imagery). There will also be an increased number of spectral channels (from 12 to 16 channels) with enhanced capability to observe cloud patterns, cloud microphysics, aerosols, and wildfires. The spatial resolution will also increase, ranging between 500 m and 1 km. Further, there will be added lightning imaging capability from the MTG Lightning Imager (LI) in order to observe cloud-to-cloud and cloud-to-ground lightning. Better knowledge of lightning strikes and the state of electrification of thunderstorms will improve aviation flight planning, enabling pilots to avoid electrically active thunderstorms [12].

In addition, the MTG program will carry a MTG-S hyperspectral Infrared Sounder (IRS). This new capability will allow for the first time a geostationary satellite to not just image weather systems but also to deliver vertical profiles of temperature and moisture over the full Earth disk, with an every hour frequency and at 4 km resolution. This will contribute greatly to numerical weather prediction modeling and for now-casting and forecasting of severe weather hours earlier. Improved imagery and infrared sounding information can also be combined to improve accuracy of wind field information, including the altitude assigned to wind vectors, dispersion of atmospheric pollution, and aerosol monitoring.

Geostationary satellites complement observations made by polar-orbiting satellites, which provide a unique wealth of ocean, land, and atmospheric parameters that can best be measured from low-altitude orbits. The EPS-SG planned for the 2020 timeframe contributes to the JPSS program set up by NOAA, and will play an important role in improving numerical weather prediction and long-term climate monitoring [13]. Advanced instrumentation is planned for EPS-SG payloads, including multi-spectral imaging, atmospheric sounding in the optical and microwave spectral domains, radio occultation sounding, scatterometry, and microwave imaging.

Some breakthroughs of the EPS-SG program include doubling of radiometric and spectral resolution of infrared sounding, resulting in 75% more information in temperature profiling and 30% more information in water vapor profiling [14]. Visible/infrared imaging is also improved with higher spatial resolution (250–500 m) cloud products and more spectral channels for accurate quantification of climate variables. Improved instrumentation will also enhance observations of ocean surface wind vectors, ozone, and aerosols. Weather observations will be provided to users in near real-time by direct broadcast. The current meteorological-satellite spectrum in the X-band (e.g., 7,750–7,900 MHz) is being considered for this broadcast service. It is expected that the delivery of observations by the instrument to the end user will range between 60 and 120 min in frequency, depending on geographic location. The Meteosat and EPS satellite systems jointly form the pillars of EUMETSAT. These satellites will help maintain a continuity of observations into the future and will enhance vital weather and climate observations, as well as now-casting of high impact weather events.

As discussed in Chap. 6, China, India, Japan, and Russia all operate meteorological and climate-monitoring satellites. These form part of the GOS promoted by the WMO, which contributes to the GEOSS coordinated by the GEO. Similar to the United States and Europe, future development of these meteorological satellite programs have moved towards enhancing now-casting abilities and producing high-resolution products for supporting weather monitoring, severe weather warning, and disaster mitigation. For example, the next generation Chinese FY-3 polar-orbiting meteorological satellite series is expected to provide substantial improvements over its FY-1 spacecraft series. The FY-3 series is to provide global air temperature, humidity profiles, and meteorological parameters, such as cloud and surface radiation required for weather forecasts, especially for medium numerical forecasting [15]. This is essential for observing climate variability and monitoring large-scale weather-induced hazards and environmental changes, providing meteorological information for aviation, navigation, agriculture, forestry, hydrology, and other economic sectors.

The latest weather satellite mission developments of Russia are quite similar to those for the United States' NOAA/NASA and Europe's EUMETSAT/ESA. The latest geostationary Elektro-L series with full-disk imaging capabilities and a wide-angle special MSU-GS scanner (with 20-degree angle of view) enables global weather forecasting, analysis of oceanic conditions, and space weather monitoring. Unlike NOAA/NASA GOES satellites, Elektro-L captures images in the infrared, as well as the visible spectrum. The near infrared channel of Elektro-L's 10-channel MSU-GS imagery has been useful not only for monitoring cloud movement but also for vegetation dynamics [16]. The spatial resolution of the infrared and visible channels is 1 and 4 km, respectively, capable of producing images every 15 to 30 min. Similar instruments are planned for the follow-up Meteor-M polar-orbiting satellite series.

Japan plans to launch its latest satellite, Himawari-8, in summer 2014 with operations commencing in 2015, along with the launch of Himawari-9 in 2016. Planned functions and specifications will be notably improved from the previous MTSAT imagery. These new features will enable better now-casting, improved numerical weather prediction accuracy, and enhanced environmental monitoring capabilities [17]. There will be higher spatial resolution data (0.5–1 km), more spectral bands (16 bands), and more frequent observations up to every 2.5 min around Japan, and also covering the east Asia and western Pacific regions (see Fig. 8.4).

Similarly, India has also greatly enhanced its INSAT geostationary meteorological satellite series [19]. Its latest INSAT-3 series incorporates state-of-the-art communication technology and significant technological improvements in sensor capabilities from its earlier INSAT missions, including a 6-channel imager and 19-channel sounder. This enables derivation of vertical temperature and moisture profiles with a resolution of 10 km and full disk coverage every half hour with the aim of improving understanding of meso-scale systems and monitoring of land and ocean surfaces for weather forecasting and disaster warning. These developments have led to a cooperation agreement between India and the U.S. to expand on INSAT-3D imager and sounder data for a variety of scientific applications, such as rainfall estimation, sea surface temperature retrieval, vegetation index retrieval, cloud classification, cloud motion vector, and water vapor winds [20]. A similar agreement exists between India and France with the Megha-Tropiques experimental mission, with

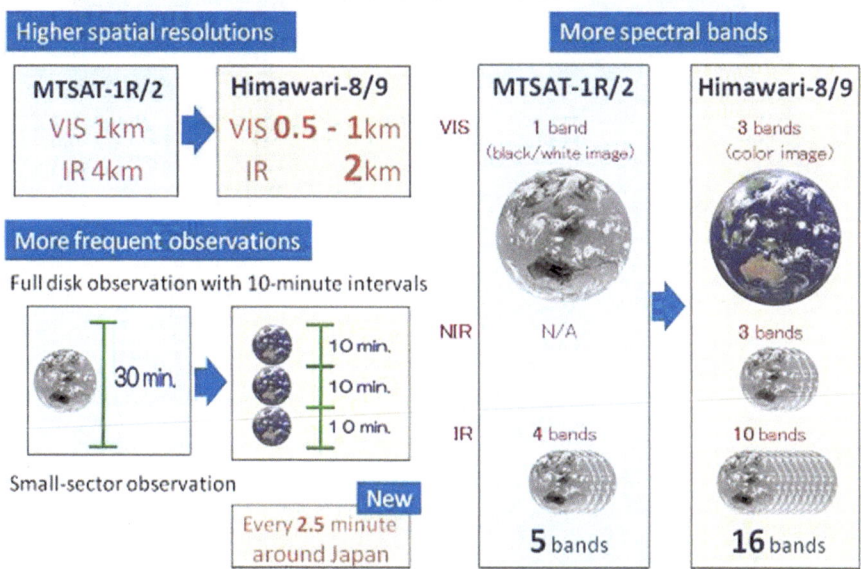

**Fig. 8.4** Improvements of the next generation Japanese Himawari-8/9 satellite as compared to that of MTSAT-1R/2. (Courtesy of Japan Meteorological Agency, JMA) [18]

the objective of studying the water cycle and energy exchanges in the tropical belt. Megha-Tropiques has a unique combination of scientific payloads, including the MADRAS instrument for sensing cloud properties and precipitation, the SAPHIR instrument for water vapor profiles, and the ScaRaB for capturing radiative fluxes. These features were discussed earlier in more detail in Chap. 6.

Collectively, weather satellites flown by the United States, Europe, India, China, Russia, and Japan provide nearly continuous observations for a global weather watch. These weather satellites have also contributed to other initiatives, such as the International COSPAS-SARSAT program, which is an international, humanitarian, satellite-based search and rescue (SAR) distress alert detection and information distribution system, originally established by the former U.S.S.R., Canada, France, and the United States in 1979 [21]. Satellite payloads are part of an international cooperative satellite-based radiolocation system to support SAR operations, which consists of both a ground segment and a space segment (see Fig. 8.5). As of 2013, 43 countries and organizations participate in the operation and management of the system. This number includes the 4 parties to the Memorandum of Agreement (Canada, France, Russia, and the U.S.), 26 ground segment providers, 11 user states, and 2 organizations.

In particular, the space segment of COSPAS-SARSAT consists of satellites in low-altitude Earth orbit (the LEOSAR System) and geostationary Earth orbit (the GEOSAR System). The LEOSAR System currently consists of NOAA POES and EUMETSAT MetOp-A satellites. The GEOSAR satellites provide continuous coverage of Earth below 70 degrees latitude with five geostationary satellites, including

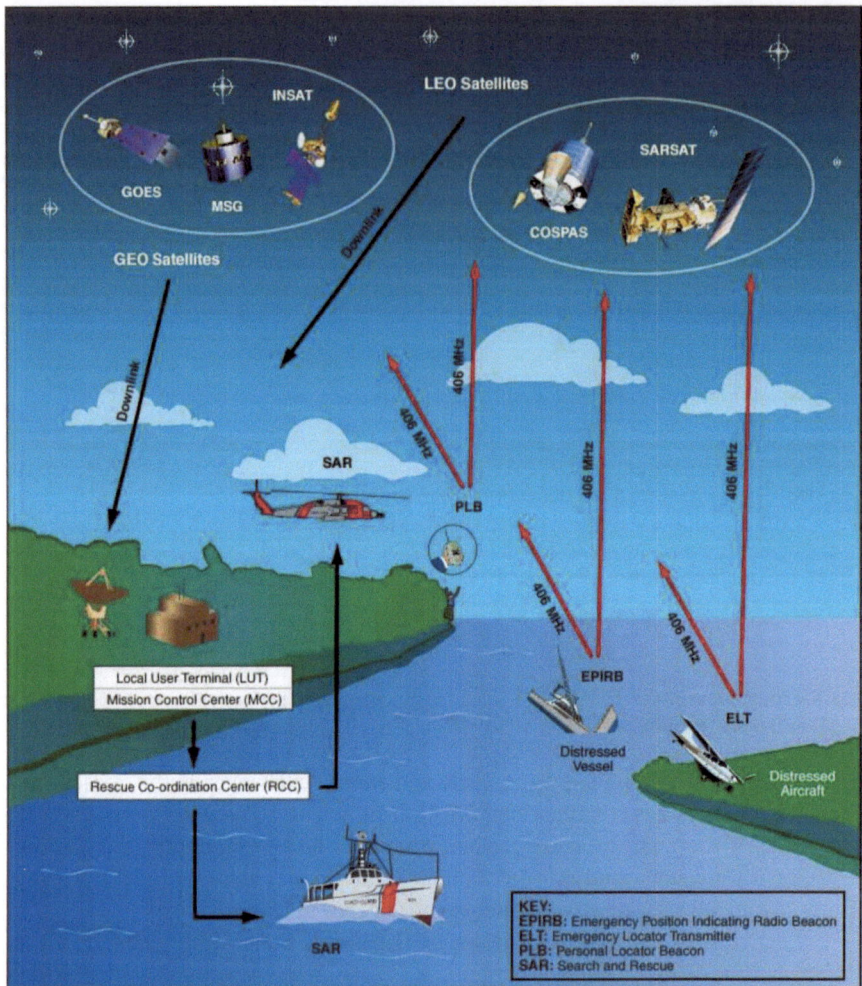

**Fig. 8.5** Overall system configuration of COSPAS-SARSAT, including both ground and space segments. (Courtesy of COSPAS-SARSAT) [22]

NOAA's GOES satellites (GOES-13 and -15), India's INSAT-3A, Europe's MSG satellites (MSG-1 and -2), and Russia's Electro-L-1 satellite. All satellites carry a payload specifically to provide services for relaying signals from distress beacons in sea, air, or land for the COSPAS-SARSAT system. Use of the system is free to the beacon operator and once rescue signals are detected and verified, search and rescue operations can be initiated, greatly supporting aircraft, ships, and land travelers in distress.

In the near future, there will be new series of both polar-orbiting and geostationary meteorological satellites developed and launched internationally with improved versions of current instruments or entirely new instruments onboard. As discussed, the general trend is towards sensing in additional areas of the electromagnetic spectrum with newly developed multi-spectral and hyperspectral instruments, providing higher resolution data at higher data transmission rates. Concurrently, the world's meteorological organizations are moving towards more international cooperation with further integration of systems and data sharing agreements. With the development of GEOSS, additional information, tools, services, and data sources will be available to the direct data user. These factors combined will result in vastly increased amounts of data being transmitted from satellites, at higher transmission rates, and with further use of data compression techniques to transmit the data within available bandwidth.

There is also a steady and continued transition from analog Automatic Picture Transmission (APT) to all digital transmission services for direct readout from meteorological satellites to ground stations. The MetOp program has switched to Low-Rate Picture Transmission (LRPT) for its new polar-orbit satellites, while NOAA has stated that all analog APT transmitters will be replaced on POES vehicles after NOAA-N [23]. LRPT is a digital transmission system intended to deliver images and data from an orbital weather satellite directly to end users via a Very High Frequency (VHF) radio signal, which is not compatible with APT receive systems that have been used since the 1960s. The LRPT provides three image channels at full sensor resolution (10-bit, 1 km/pixel, six lines/second), whereas the analog APT system provides only two image channels at reduced accuracy and resolution (8-bit, 4 km/pixel, two lines/second). LRPT images are four times more accurate and contain twelve times the resolution, as well as delivering data from other sensors, such as atmospheric sounders and GPS positioning information. There will be a considerable period of overlap when MetOp, POES, and NPOESS satellites will be orbiting simultaneously to enable APT users to transition their ground stations to LRPT [24].

The planned development of GEOSS, with efforts coordinated by the intergovernmental GEO on the basis of a 10-year implementation plan for 2005–2015, will meet the need for timely, high quality, long-term global information required for decision making and enhancing understanding of the Earth system [25]. Access to environmental data will be greatly enhanced to benefit society in identified key areas of agriculture, weather, water resources, energy, health, climate, biodiversity, disaster mitigation, and ecosystems. GEONETCast is a milestone in the emerging GEOSS, designed as a data distribution system, providing environmental data to a worldwide user community [26]. Current partners include the WMO, NOAA in the United States, the CMA in China, and the European organization EUMETSAT, as well as many prospective data provider partners. Thus GEONETCast will broadly enable environmental data exchange and data delivery available in Europe, Africa, and the Americas. Additional data exchange is currently being established in the Asia Pacific region.

GEONETCast is a task in the GEO Work Plan that establishes a global data distribution system. The key is a small number of regional but interconnected GEONETCast centers that take on the responsibilities for establishing a satellite-based regional dissemination system, based upon Digital Video Broadcast (DVB)

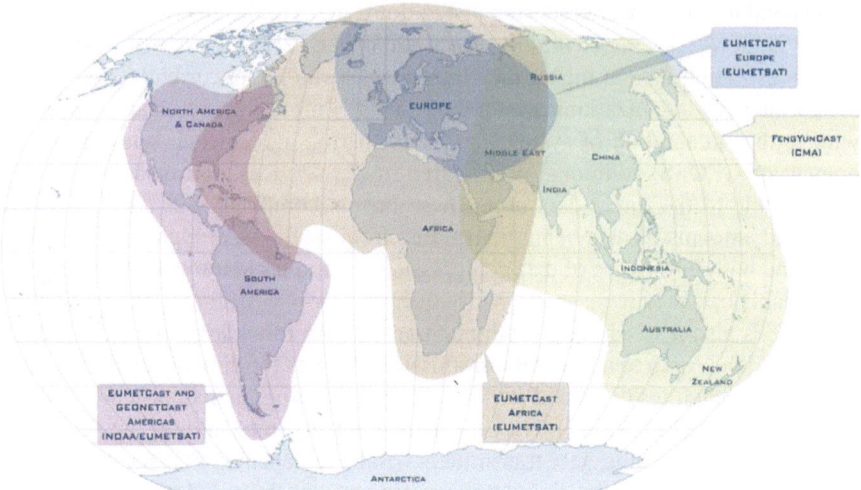

**Fig. 8.6** GEONETCast regional broadcast footprints. (Courtesy of GEO) [27]

technology similar to that used in satellite television. This is especially useful in parts of the world where high speed land lines and/or internet are not available, or in regions where terrestrial communication lines have been disrupted by disasters. GEONETCast is currently implemented in Europe by EUMETCast with coverage over Europe, Africa, and the Americas. FengYunCast is the Chinese contribution providing coverage of the Asia Pacific region, while GEONETCast Americas provides broadcast covering the Americas and managed by NOAA (see Fig. 8.6). Products include diverse raw data, as well as processed value-added products from observing data platforms, including operational or research-based remote-sensing systems such as satellites and ground-based or airborne platforms.

Digital Video Broadcast (DVB) technology enables broadcast services to be received by relatively low-cost user receiver stations, which are particularly useful for developing countries, where funds may be insufficient to provide sufficient internet capacity for high volume download of Earth observation products. Other advantages of GEONETCast include the ability to use low cost, off-the-shelf receiver equipment, highly scalable system architecture, high reliability and data transfer rate. All of these features combine to provide low cost access to a wide variety of freely available products and a constantly growing network of receivers, providers, and products for weather, climate, and environmental data and information.

The GEONETCast system makes available a one-stop-shop delivery mechanism allowing users to receive many data streams via one reception station, and access to a user base of over 3,000 user stations worldwide. Available data streams typically include space-based observations from Meteosat, MetOp, NOAA, Jason-2, GOES, FY-2, and FY-3A satellites. In most cases, images are delivered to users within five minutes of processing. Available products include marine meteorological and ocean surface products, land applications products, and atmospheric chemistry products.

Such developments allow greater accessibility to environmental data in near real-time with global coverage to a wide user community. The goal of such initiatives is to enhance dissemination, application, and exploitation of environmental data and products for the diverse societal benefits defined by the GEO, including severe weather forecasting and disaster management for reducing loss of life and property.

Through GEOSS, global partners will share observations and products that are accessible, comparable, and supported by common standards suited for user needs. GEOSS will promote integration of data sources from GEO geostationary and polar-orbiting satellites, and capacity building in Earth observation that build on existing local, national, regional, and international initiatives. Collective action for global observing systems has many advantages. Such systems can better meet the needs of sustainable development, improved monitoring of the state of Earth, and enhanced prediction of Earth system behavior. All of these innovations are critical to combating the problems associated with climate change.

# References

1. WMO: http://www.wmo.int/pages/prog/amp/pwsp/Nowcasting.htm. Accessed 12 Aug 2013
2. EUMETSAT: http://www.eumetsat.int/website/home/Satellites/CurrentSatellites/Meteosat/RapidScanningService/index.html. Accessed 12 Aug 2013
3. NOAA: Nowcasting the Wind Speed during a Hurricane at Sea. Mariners Weather Log. Spring/Summer **47**(1) (2003)
4. NOAASIS: http://www.noaasis.noaa.gov/NOAASIS/ml/future.html. Accessed 12 Aug 2013
5. GOES-R: http://www.goes-r.gov/spacesegment/glm.html. Accessed 12 Aug 2013
6. Demaria, M., Demaria, R. T., Knaff, J. A., Molenar, D.: Tropical Cyclone Lightning and Rapid Intensity Change. Monthly Weather Review, **140**, 1828–1842 (2012)
7. Bloom, H. J.: Next Generation Geostationary Operational Environmental Satellite: GOES-R, the United States' Advanced Weather Sentinel. Proceedings of SPIE, vol. 7458 745802–1 (2009)
8. GOES-R: http://www.goes-r.gov/spacesegment/glm.html. Accessed 12 Aug 2013
9. GOES-R: http://www.goes-r.gov/spacesegment/glm-lightning-detect.html. Accessed 12 Aug 2013
10. NOAA: http://www.jpss.noaa.gov. Accessed 12 Aug 2013
11. EUMETSAT: Meteosat Third Generation: Europe's Future Geostationary Meteorological Satellite System. Brochure PRG.FS.01.V.2. (2012)
12. EUMETSAT: http://www.eumetsat.int/website/home/Satellites/FutureSatellites/MeteosatThirdGeneration/MTGDesign/index.html. Accessed 12 Aug 2013
13. EUMETSAT: http://www.eumetsat.int/website/home/Satellites/FutureSatellites/EUMETSATPolarSystemSecondGeneration/index.html. Accessed 12 Aug 2013
14. Schlussel, P.: EUMETSAT Polar System – Second Generation. Presentation, EUM/MET/VWG/12/0160, Issue 1 (2012)
15. eoPortal: https://directory.eoportal.org/web/eoportal/satellite-missions/f/fy-3. Accessed 12 Aug 2013
16. eoPortal: https://directory.eoportal.org/web/eoportal/satellite-missions/e/electro-l. Accessed 12 Aug 2013
17. JMA: http://mscweb.kishou.go.jp/himawari89/index.html. Accessed 12 Aug 2013
18. JMA: http://mscweb.kishou.go.jp/himawari89/index.html. Accessed 12 Aug 2013
19. Singh, D., Nair, S.: Current Status and Future Prospects of Indian Satellite. Presentation, Department of Science and Technology, New Delhi (2007)

20. eoPortal: https://directory.eoportal.org/web/eoportal/satellite-missions/i/insat-3. Accessed 12 Aug 2013
21. COSPAS-SARSAT: http://www.cospas-sarsat.org. Accessed 12 Aug 2013
22. COSPAS-SARSAT: http://www.cospas-sarsat.org. Accessed 12 Aug 2013
23. Kramer, H. J.: Observation of the Earth and Its Environment: Survey of Missions and Sensors, 4th ed. Springer, New York (2002)
24. U.S. Department of Commerce, NOAA, and NESDIS: User's Guide for Building and Operating Environmental Satellite Receiving Stations. http://www.noaasis.noaa.gov/NOAASIS/pubs/Users_Guide-Building_Receive_Stations_March_2009.pdf (2009). Accessed 12 Aug 2013
25. GEO: The Global Earth Observation System of Systems (GEOSS) 10-Year Implementation Plan. GEO 1000R: February 2005. ESA Publications, Noordwijk (2005)
26. GEO: http://www.earthobservations.org/geonetcast.shtml. Accessed 12 Aug 2013
27. GEO: http://www.earthobservations.org/geonetcast.shtml. Accessed 12 Aug 2013

# Chapter 9
# Meteorological and Remote-Sensing Satellites in Monitoring Climate Change

> *Global observations coordinated by WMO show that levels of carbon dioxide, the most abundant greenhouse gas in the atmosphere continue to increase steadily and show no signs of leveling off.*
>
> —Michel Jarraud, Secretary-General of the WMO

With the continued evolution of satellite instruments and technology, meteorological satellites enable scientists to track long-term and short-term variability in climate. Such observations contribute to our understanding of climate change and its impacts. This is necessary for governments and decision-makers to define and implement appropriate mitigation and adaptation policies, and to provide new environmental services. Accurate long-term climate monitoring and predictions start from the best possible knowledge of the state of Earth's various systems, including oceans, land, atmosphere, biosphere, and cryosphere. Satellite spacecraft and instruments enable continuous observation of solar activity, sea level rise, the temperature of the atmosphere and oceans, the state of the ozone layer, air pollution, and changes in sea ice and land ice. 'National needs' for weather forecasts and climate monitoring are turning into global needs that transcend political boundaries. Space instruments also allow the monitoring and gathering of data on space weather and solar activity that can greatly impact life on our planet.

Scientists have high confidence that global temperatures will continue to rise for decades to come, largely due to greenhouse gases produced by human activities. Nearly 30 years of satellite-based solar and atmospheric temperature data have assisted the Intergovernmental Panel on Climate Change (IPCC) to conclude in 2007 that "most of the observed increase in global average temperatures since the mid-20th century is *very likely* due to the observed increase in anthropogenic greenhouse gas concentrations." [1] Moreover, the IPCC predicts that global temperatures will rise 2.5–10 degrees Fahrenheit over the next century and that "taken as a whole, the range of published evidence indicates that the net damage costs of climate change are likely to be significant and to increase over time" [2].

However, critical questions still remain as to what the consequences of global warming and climate change will be. For example, how much warmer will global temperatures rise? How much will sea levels increase? What will happen to soil moisture on a warmer planet and how will this impact agriculture? How will this affect extreme weather conditions? Observations by operational meteorological

S.-Y. Tan, *Meteorological Satellite Systems*, SpringerBriefs in Space Development, DOI 10.1007/978-1-4614-9420-1_9, © The author 2014 satellites will play an important role in providing data and information to support climate monitoring, climate change analysis and modeling, and other climate-related products and services.

## Global Climate Observing System (GCOS) Essential Climate Variables

As previously discussed in Chap. 7, the GCOS was established in 1992 to ensure that the observations and information needed to address climate-related issues are obtained in a timely way and then made available to all potential users. GCOS is a long-term, user-driven operational system to provide the comprehensive observations required for monitoring the climate system, for detecting and attributing climate change, for assessing the impacts of climate variability and change, and for supporting research toward improved understanding and prediction of the climate system. It addresses the total climate system, including physical, chemical and biological properties, and atmospheric, oceanic, terrestrial, and cryospheric processes. It is co-sponsored by the WMO, the Intergovernmental Oceanographic Commission (IOC) of UNESCO, the United Nations Environment Programme (UNEP), and the International Council for Science (ICSU).

GCOS has defined a list of 50 ECVs that are both currently technically and economically feasible for global studies and have a high impact on the requirements of the United Nations Framework Convention on Climate Change (UNFCC) and the IPCC [3]. This list is not exhaustive. Other climate variables may be important for fully understanding the climate system, but these may be part of ongoing research and may not be ready for global studies on a systematic basis. The GCOS ECVs are currently measurable and require international exchange of both current and historical observations for monitoring physical, chemical, and biological components of Earth.

ECVs help scientists to effectively keep track of how Earth is changing in accordance with a standardized measuring system established by GCOS. These are effectively the planet's "vital signs," which include surface air temperatures, upper air temperatures, atmospheric composition (ozone levels, carbon dioxide and aerosol properties), ocean surface temperatures, snow cover, ice cap cover, river discharge, soil moisture, etc. While some variables are useful for longer-term climate predictions (i.e., radiation budget, sea surface salinity), other variables are useful for short-term weather predications, such as air pressure and upper air wind speed and direction. ECVs are useful for many reasons. These include air quality monitoring and forecasts, water resource management, and even for assessing the likelihood of disease outbreaks. The chart in Fig. 9.1 shows the complexity of the various measurements.

GCOS is part of the larger effort to build GEOSS, which is an international plan to support the work of the IPCC for comprehensive, coordinated, and sustained Earth observation (refer to Chap. 7). Since most ECVs can be measured from space,

| The Essential Climate Variables | | |
|---|---|---|
| **Domain** | **Essential Climate Variables** | |
| **Atmospheric** (over land, sea and ice) | Surface: | Air temperature, **precipitation**, air pressure, surface radiation budget, wind speed and direction, water vapour. |
| | Upper air: | Earth radiation budget (including solar irradiance), upper air temperature (including MSU radiances), wind speed and direction, water vapour, cloud properties. |
| | Composition: | **Carbon dioxide**, methane, **ozone**, other long-lived greenhouse gases, **aerosol properties**. |
| **Oceanic** | Surface: | **Sea surface temperature**, sea surface salinity, **sea level, sea state, sea ice**, currents, **ocean colour (for biological activity)**, carbon dioxide partial pressure. |
| | Sub-surface: | Temperature, **salinity**, currents, nutrients, carbon, ocean tracers, phytoplankton. |
| **Terrestrial** | River discharge, water use, ground water, **lake levels, snow cover, glaciers and ice caps**, permafrost and seasonally-frozen ground, **albedo, land cover** (including vegetation type), **fraction of absorbed photosynthetically active radiation (fAPAR)**, **leaf area index (LAI), biomass, fire disturbance, soil moisture**. | |

**Fig. 9.1** The GCOS ECVs categorized based on three domains of atmospheric, oceanic, and terrestrial observations. Variables in bold type are largely dependent on satellite observations. (Courtesy of CEOS)[4]

satellite-based monitoring and Earth observation form a key part of the GEOSS effort. Moreover, a comprehensive global climate observing system requires observations not only from satellites but also observations from land-based, airborne *in situ*, and remote-sensing platforms. Data from instruments at ground stations, ships, buoys, ocean profilers, balloons, samplers, aircraft, and satellites are transformed into useful environmental products through analysis and integration. Such Earth observation products provide the necessary evidence for informed environmental decision-making in support of vital climate studies.

As climate monitoring becomes an increasingly important worldwide concern, satellites are expected to become widely used as a means of obtaining both regional and global observations of ECVs. Meteorological satellites provide long-term observations of Earth and the space environment from which environmental and meteorological variables can be derived. However, some ECVs still remain dependent on *in situ* observations for long-term trend information, for calibration and validation of satellite records (sometimes known as "ground-truthing"), and for measuring variables not amenable to direct satellite measurement (e.g., sub-surface oceanic ECVs) [5]. Although most ECVs can be measured from space and satellite platforms, technical limitations require that some parameters, such as river discharge, be monitored at the surface and from *in situ* measurements. Moreover, ECVs assist in defining and updating climate requirements for global observing systems, as well as highlighting gaps in future plans and capabilities for more effective climate monitoring.

The benefits of space-based Earth observations satisfy a broad range of users, including national, regional, and local decision-makers. With more than 30 years of consistent data, satellites have provided unprecedented data sources and continuous support for long-term and short-term climate prediction and climate monitoring capability. Meteorological satellites provide an invaluable data source for understanding our changing climate, and their role will grow increasingly important with coming generations of remote-sensing systems and sensor technologies. The following sections provide an overview of current space-based monitoring systems for selected ECVs.

## Atmospheric Ozone

Over the coming decades, high quality global observations of ozone and ozone-depleting substances would be particularly critical in verifying the effectiveness of the actions taken under the Vienna Convention in 1985, the Montreal Protocol of 1987 and its amendments. (Michel Jarraud, Secretary-General of WMO).

With the continued evolution of satellite instruments and technology, meteorological satellites are able to provide scientists with the ability to track long-term and short-term changes in atmospheric constituents, such as atmospheric ozone. Ozone plays an essential role in complex mid-atmospheric photochemistry. Its depletion, particularly in the so-called ozone holes in the polar regions, is a severe manifestation of climate change that can have harmful consequences for humans, plants, and animals [1]. Moreover, stratospheric ozone depletion can affect local surface temperature and trigger changes in the planet's climate.

Changes in ozone, including its diurnal, seasonal, and yearly variability can be detected and even predicted by means of satellite observations [6]. For example, TOVS consists of three instruments designed to determine radiances required to calculate temperature and humidity profiles of the atmosphere from the surface to the stratosphere. These include the HIRS, the Microwave Sounding Unit (MSU), and the Stratospheric Sounding Unit (SSU). The HIRS includes a channel 9 (that measures in the 960 nanometer—nm—wavelength) that is able to retrieve total column ozone values and to help track ozone patterns.

In Europe, the second-generation Meteosat satellite (Meteosat-8 or MSG-1) was launched in 2002, carrying a main optical payload Spinning Enhanced Visible and Infrared Imager (SEVIRI) with a 970-nm infrared channel called the "ozone channel." This is used to routinely and synoptically monitor ozone concentration in the ozone layer (i.e., lower stratosphere from 12 to 36 km) by measuring up-swelling infrared radiation in the 970-nm region. Total ozone and height of the tropopause can be monitored. Meteorological satellites in geostationary orbit, such as Meteosat, also provide full Earth disk coverage, thus enabling entire ozone reconnaissance conveys to be performed, providing a wealth of information concerning dynamical processes in the lower stratosphere [7].

Since 1994, the new generation GOES sounders have provided higher spatial and temporal resolution radiance measurements that are able to detect ozone. For

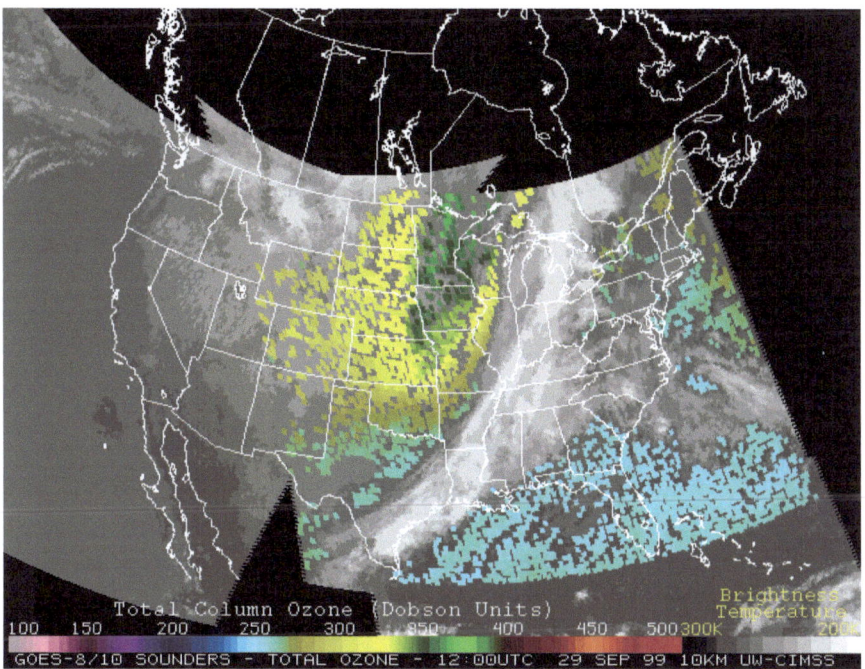

**Fig. 9.2** GOES-8 sounder total ozone product showing a lobe of high ozone concentration (350–375 Dobson units, green enhancement) over the North Central United States on September 29, 1999. (Courtesy of CIMSS) [10]

example, GOES-8 infrared observations have been used to derive high-resolution synoptic maps of total ozone over North America, which have been applied for fine-scale pattern identification and high-resolution ozone simulation (Fig. 9.2) [8].GOES-8/9 sounders measure radiances from 8-km footprints in one visible and 18 infrared spectral bands, many of these sensitive to atmospheric carbon dioxide, ozone, and water vapor [9]. Spectral bands are selected to probe the atmosphere at different depths, enabling the sensing of vertical variations of temperature, moisture, and ozone levels.

The Solar Backscatter Ultraviolet (SBUV) instrument is an operational remote sensor designed specifically to map total ozone concentrations and the vertical distribution of ozone in Earth's atmosphere on a global scale. Ozone measurements from this type of backscatter ultraviolet instrument have been made since the launch of Nimbus-4 in April 1970 with the SBUV instrument. This continued with the second generation instrument, SBUV/2, on the NOAA TIROS series of satellites in 1984, continuing with the latest NOAA-19 spacecraft launched in 2009. The instrument itself measures solar irradiance and Earth radiance in the near ultraviolet spectrum (160–400 nm) (Fig. 9.3).

The Total Ozone Mapping Spectrometer (TOMS) developed by NASA also measures ozone concentrations, as well as tropospheric aerosols, volcanic sulfur

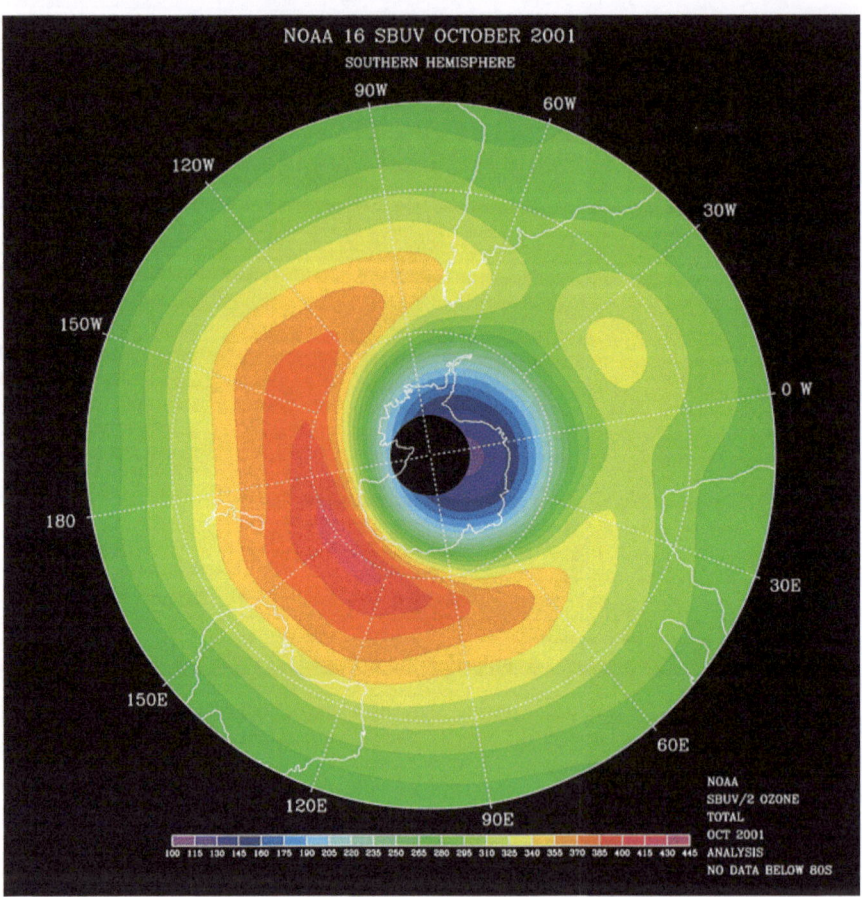

**Fig. 9.3** A false color contour map of the total atmospheric ozone for October 2001 over the southern hemisphere measured by the NOAA-16 SBUV/2 instrument. (Courtesy of NESDIS) [11]

dioxide, and ultraviolet irradiance among other physical parameters [12]. Four TOMS instruments have been successfully flown in orbit aboard the Nimbus-7 (1978–1993), Meteor-3 (1991–1994), Earth Probe (1996-present), and ADEOS (1996–1997) satellites. They are able to provide global measurements of total column ozone on a daily basis with high-resolution observations covering the near ultraviolet region. Ever since 2006, the Ozone Monitoring Instrument (OMI) flown onboard the NASA Aura satellite has replaced the TOMS and SBUV sensors.

The Ozone Mapping Profiler Suite (OMPS) is the latest generation of backscattered ultraviolet radiation sensors, which is currently flying onboard the Suomi NPP spacecraft and will also fly on future NPOESS satellites. It has a dual mission to provide NOAA with critical operational ozone measurements, as well as continuing the 40-year NASA records of total column and profile ozone created

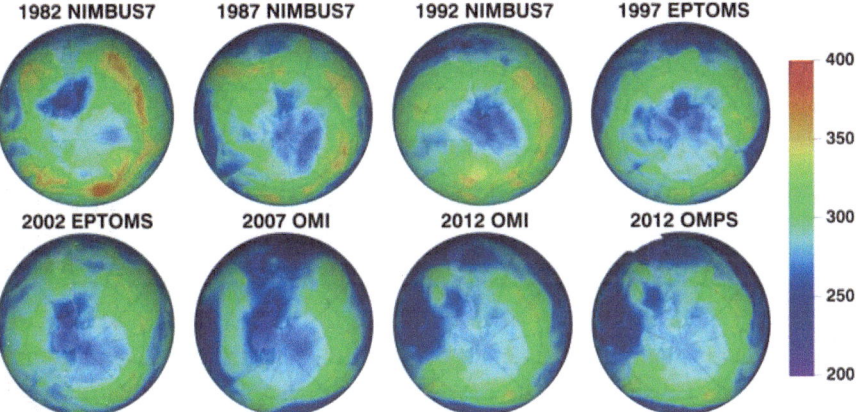

**Fig. 9.4** Tracking changes in ozone concentration on January 27 from 1982 to 2012, based on more than 30 years of ozone data. The thickness of the ozone layer is measured in Dobson units, with smaller concentrations of overhead ozone shown in blue, while larger amounts are shown in orange and yellow. (Courtesy of NASA) [14]

by previous sensors [13]. OMPS is comprised of three different sensors, two nadir instruments looking directly below the satellite, and one limb instrument looking at an angle to Earth's surface. Combining the two views enhances the ability of scientists to measure the vertical structure of ozone. OMPS will continue the U.S. program for monitoring Earth's ozone layer, building upon lessons learned from previous SBUV, SBUV/2, and TOMS instruments on NASA/NOAA and international missions. Near real-time measurement of changes in the ozone layer is now operationally possible with the availability of high temporal and spatial resolution data, enabling global ozone distribution to be mapped in detail, as well as the Antarctic "ozone hole" (Fig. 9.4).

## Precipitation and Cloud Patterns

> Think of hurricanes as a fire in a fireplace. Global warming isn't the spark, but it is that extra log we are throwing onto that fire. (Scott Mandia).

Global rain measurement plays an important role in assessing the hydrologic and agricultural potential of remote areas. These measurements are also useful in detecting excessive convective rains in remote/mountainous areas for flood warnings, and evaluating the flood potential of tropical and extra-tropical systems as they approach land. According to meteorological research, satellite data have been successfully applied to estimate both the volume and area distribution of rainfall, which are very difficult to assess using the conventional rain gauge method.

**Table 9.1** Rainfall probabilities and intensities related to dominant cloud type categories. (Reproduced from Kidd, 2011) [15]

| Cloud type category | Relative probability of rainfall | Relative intensity of rainfall |
|---|---|---|
| Cumulonimbus | 0.90 | 0.80 |
| Stratiform | 0.50 | 0.50 |
| Cumuliform | 0.10 | 0.20 |
| Stratocumuliform | 0.10 | 0.01 |
| Cirriform | 0.10 | 0.01 |
| Clear skies | – | |

Although no meteorological satellite yet exists that can reliably identify rainfall and accurately estimate rainfall rate in all circumstances, satellites carry sensors that can make indirect estimates of rainfall by taking measurements of cloud thickness or temperature.

Meteorological satellites since TIROS-1 have been applied for detecting cloud patterns and cloud systems. Satellite systems have advanced markedly since then, but often their spatial resolution and poor temporal sampling has been inadequate for rainfall applications [15]. Polar-orbiting satellites carrying the AVHRR sensor, which senses in the visible and infrared spectra, are able to generate high-resolution imagery for global temperature and precipitation measurements, but poor temporal viewing (i.e., every 6 hours). This lack of up-to-date information has proven to be insufficient for some meteorological applications [16].

In contrast, geostationary satellites are able to continuously observe cloud patterns, while visible/infrared imagery can be used to distinguish rain-filled clouds from non-rain filled clouds on the basis of their observed cloud top temperature. This is on the premise that rain originates from deep convective clouds with cold, high tops, such as cumulonimbus. Furthermore high, wispy clouds called cirrus tend to form after storms and occur high in the atmosphere. These clouds are mainly composed of ice crystals, rather than water drops; thus this type of cloud appears as very cold in satellite imagery. Rainfall probability and intensity can be determined by cloud types recognized from satellite imagery. Table 9.1 shows rainfall estimates derived from visible/infrared data based on a cloud indexing method, where intensity values are assigned to different cloud types in an area. These techniques use visible imagery for identifying cloud type, cloud area, and cloud growth rate observations, whereas infrared imagery provide information about cloud top temperature and growth rate.

Another technique for rainfall estimation is the life-history approach, which also uses visible/infrared geostationary imagery but depends on good temporal resolution data for tracking cloud development and movement. This has been applied to estimate convective rainfall from GOES data and to provide an every 3-hour outlook of rainfall estimates [17]. This technique relies on identifying thunderstorm clusters and assessing their temperature and past development in order to adjust rainfall estimates and forecasts.

Arguably the most widely used technique is the GOES Precipitation Index (GPI). This index allows the derivation of an optimum cold cloud threshold with an associated rainfall intensity [18]. The basis for this group of assessment techniques is that the rainfall intensities will be different according to the life cycle stage of the convective cloud.

There have been some initiatives to combine precipitation information available from several sources into a final merged product, taking advantage of the strengths of each data type. The Global Precipitation Climatology Project (GPCP) combines microwave estimates from the Special Sensor Microwave/Imagery (SSM/I) data from the DMSP, along with infrared data obtained primarily from geostationary satellites (United States, Europe, Japan). Additional data is secondarily obtained from American polar-orbiting satellites, along with over 6,000 rain gauge stations [19]. GPCP data have been used to reveal changes in observed precipitation on seasonal to yearly time scales. This type of data offers the potential for studying changes in precipitation at longer time scales, especially over oceans. Similar initiatives that produce global precipitation products include the Climate Prediction Center Merged Analysis of Precipitation (CMAP) [20] and the CPC MORPHing technique (CMORPH) [21].

The Tropical Rainfall Measurement Mission (TRMM) was launched in 1997 and is being flown by NASA and JAXA to improve our quantitative knowledge of the three-dimensional distribution of precipitation in the tropics. It is the first on-orbit active/passive instrument package to study the intensity and structure of tropical rainfall [22]. TRMM has a passive microwave radiometer, the first active spaceborne Precipitation Radar (PR), and a Visible-Infrared Scanner (VIRS), Clouds and the Earth's Radiant Energy Sensor (CERES), and Lightning Imaging Sensor (LIS) along with other instruments. Products include the TRMM and Other Satellites precipitation estimate and the TRMM and Other Sources precipitation estimate. It optimally merges microwave and infrared rain estimates to produce on a three-hour timetable updated precipitation fields at quarter degree spatial resolution. This data is then aggregated and merged with rain gauge data to produce the best-estimated monthly precipitation field. An example TRMM dataset is shown in Fig. 9.5.

Building upon TRMM's legacy, the next-generation Global Precipitation Measurement (GPM) mission is to be launched by NASA and JAXA in 2014. This new satellite system capability will set new standards for precipitation and snow measurements worldwide. GPM will use a network of satellites united by a "GPM Core Observatory" satellite, which will carry a dual-frequency precipitation radar and microwave imager (Fig. 9.6) [24]. These sensors will provide information on precipitation particles, layer-by-layer, and within clouds. The passive microwave imager will be able to sense the total precipitation within all cloud layers. GPM instruments will extend the capabilities of TRMM sensors, including sensing falling snow, light rain, and, for the first time, quantitative estimates of microphysical properties of precipitation particles. The GPM mission will help improve the accuracy of weather forecasts and extend current satellite precipitation monitoring capabilities.

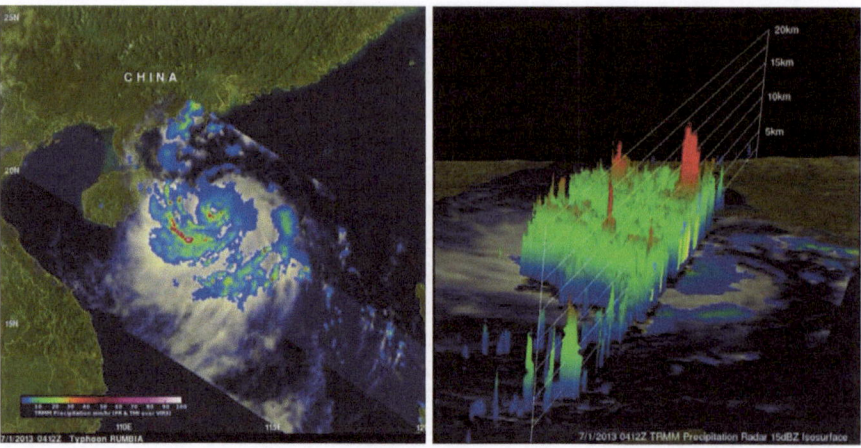

**Fig. 9.5** TRIMM satellite data from typhoon Rumbia in the South China Sea on July 1, 2013. The 3-D image on the right is derived from TRIMM's Precipitation Radar (PR) and shows the eye of the storm reaching heights of 17 km. (Courtesy of NASA) [23]

**Fig. 9.6** Visualization of the Global Precipitation Measurement (GPM) Core Observatory and partner satellites. (Courtesy of NASA) [24]

## Ocean Dynamics

El Niño and La Niña are long-term natural oceanic phenomena associated with atmospheric interactions over the equatorial Pacific Ocean every 3–5 years [25]. During these events, sea surface temperature changes over the Pacific Ocean, subsequently affecting climate, weather, atmospheric winds, and rainfall patterns around the globe. El Niño is characterized by unusually warm temperatures and La Niña by unusually cool temperatures in the equatorial Pacific. Southern Oscillation refers to an atmospheric phenomenon that occurs when an atmospheric pressure fluctuates, depending on the variations of the sea surface temperature of the Pacific Ocean. When an oceanic phenomenon such as El Niño/La Niña is coupled with

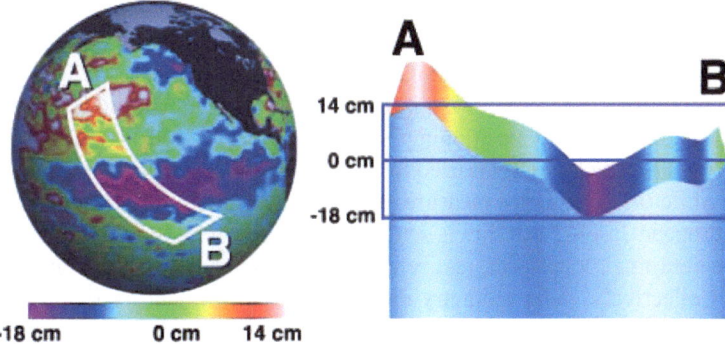

**Fig. 9.7** Ocean surface topography. Colors represent higher or lower than normal sea surface height. (Courtesy of NASA) [28]

this Southern Oscillation, it is called the El Niño/La Niña—Southern Oscillation (ENSO) event. This represents yearly variability of global-scale circulation associated with atmosphere-ocean coupling.

Since the 1970s, El Niño and La Niña events have been occurring with more frequency and intensity due to global warning. Such climate pattern oscillations have caused intensification of extreme weather (such as floods and droughts) in many regions of the world, particularly those bordering the Pacific Ocean. Since 1982, NOAA has collected sea surface temperature and cloud pattern observations from the AVHRR sensor carried by NOAA polar-orbiting meteorological satellites. Measurements from the SSM/I sensor onboard DMSP satellites have been particularly valuable for studying precipitation variations associated with ENSO events [26].

ENSO events can be seen in measurements of sea surface temperature, which can be observed in both infrared and microwave regions [27]. Most importantly, Ocean Surface Topography (OST) is an important parameter for detecting El Niño and La Niña phenomena. These OST parameters can help gather long-term information about the world's ocean and its currents (Fig. 9.7). Ocean surface topography often corresponds to the amount of heat stored in the upper layers of the ocean. For example, sea level in the eastern Pacific will be higher than average during an El Niño event (i.e., warm water over the central eastern Pacific), while the western Pacific will be higher than average during a La Niña event (i.e., warm water over the western Pacific). Therefore, accurate measurements of ocean surface topography are important for studying ocean tides, circulation, and the amount of heat the ocean holds.

NASA currently operates two satellite missions that measure ocean surface topography, including Jason-1, launched in 2001, that is partnered with the French space agency, CNES. This satellite continues ocean surface measurements made by TOPEX/Poseidon, which was the first major oceanographic research satellite to be launched into space and operated from 1992 to 2006. The second mission is the follow-up Ocean Surface Topography Mission on Jason-2 (OSTM/Jason-2). This satellite was launched in 2008 and represents a partnership with NASA, NOAA,

CNES, and EUMETSAT. Both Jason-1 and Jason-2 missions use radar altimeter systems designed to make accurate and precise measurements of the ocean surface within an accuracy of about 3 cm relative to the center of Earth.

Jason-2 carries the Poseidon-3 altimeter, the Advanced Microwave Radiometer (AMR), and the Global Positioning System Payload (GPSP) to allow precision measurements. The altimeter is used for measuring sea level, wave height, and wind speed, while the radiometer measures water vapor content in the atmosphere. The U.S./European Jason-3 satellite is scheduled for launch in 2015 and is expected to provide continual, long-term, and reliable data of change in ocean surface topography. Collectively, radar altimetry data combined with data from other satellites and *in situ* measurements from ships, profiling buoys, and drifters help to improve forecasting of climatic events, such as El Niño and El Niña.

## Sea Ice

Sea ice is found in the polar regions of the Arctic and Antarctica and has a crucial role in the global climate system. Sea ice has a bright surface reflectivity (i.e., high albedo), which affects the amount of sunlight absorbed by the Earth system. Positive feedbacks associated with sea ice can cause an amplification of climate change in the Arctic. Warming causes less ice and thus less sunlight reflected back to space. This leads to enhanced warming. Sea ice also insulates the ocean from the cold atmosphere, influencing heat fluxes to the atmosphere and affecting local clouds, precipitation, and ocean circulation. Even a small increase in temperature can lead to greater warming and melting of sea ice over time, hence making the polar regions the most sensitive areas to climate change on Earth [29].

Satellite observations show that Arctic sea ice concentrations have significantly decreased over the last 25 years at a rate of approximately 6% per decade for summer ice [30]. The year 2002 marked the lowest summer ice cover on record. Passive microwave sensors on satellites have been used for measuring the spatial extent and concentration of sea ice since the 1970s. Microwave and active sensors are beneficial, as they can operate all-year round, while visible and near-infrared sensors are limited by cloud cover and available sunlight. Microwave emission is not as strongly tied to the temperature of an object and is more sensitive to the object's physical properties (i.e., structure and composition).

Sea ice observations first began in December 1972 with the Electrically Scanning Microwave Radiometer (ESMR) on NOAA's Nimbus-5 satellite. The longest continuous sea ice time series has been obtained by passive microwave sensors since 1978 with NASA's Scanning Multi-channel Microwave Radiometer (SMMR) on-board Nimbus-7. SSM/I and Special Sensor Microwave Image/Sounder (SSMIS) instruments on-board DMSP satellites, and Advanced Microwave Scanning Radiometer (AMSR) on-board the Advanced Earth Observing Satellite II (ADEOS-II) have successfully retrieved quantitatively reliable information on sea ice. Another instrument called the Advanced Microwave Scanning Radiometer for Earth

Observing System (AMSR-E) has been used to distinguish between sea ice and oceanic water, measuring sea ice extent and ice concentration (percentage of ice and open water). It was designed by JAXA and carried on-board the Aqua satellite launched by NASA in 2002.

Active microwave satellite sensors have been useful for determining physical properties of sea ice, such as ice thickness. Imaging radar instruments such as the Synthetic Aperture Radar (SAR) have been applied for distinguishing between young (thin) and old (thick) sea ice, since microwave radiation reflected back to the sensor depends on the physical properties of the surface (e.g., structure and composition). The RADARSAT mission by the Canadian Space Agency is the primary SAR mission today. RADARSAT-1 was launched in 1995 and has provided detailed images of sea ice, aiding in maritime planning and navigation. This was followed by RADARSAT-2, launched in December 2007 [31]. In addition, non-imaging radar or scatterometers have been used to measure the amount of reflected energy or backscatter from Earth's surface. An example is the SeaWinds sensor aboard NASA's Quick Scatterometer (QuikSCAT), which provides ocean winds and sea ice products.

Satellite radar altimetry, which involves sending a pulse of radar energy toward Earth and measuring return echo signals, helps to determine the total thickness of sea ice, enabling open water, new ice, or multi-year ice to be distinguished. ESA's Cryostat program was designed specifically to monitor variations in the extent and thickness of polar ice. However, the CryoSat-1 spacecraft was lost in a launch failure in 2005. The replacement, CryoSat-2, was launched on April 8, 2010, providing data about polar ice caps and tracking changes in ice thickness with a resolution of about 0.5 inches. Main payload instruments include the SAR/Interferometric Radar Altimeters (SIRAL), Doppler Orbit and Radio Positioning Integration by Satellite (DORIS), and Laser Retroreflector (LRR) [32]. Valuable observations have been achieved by CryoSat-2 of ice sheet interiors, ice sheet margins, and sea ice topography, revealing major loss of Arctic sea ice during the last decade (Fig. 9.8). Radar systems that use microwaves rather than visible light are able to operate effectively at all times and "see through" cloud cover.

Finally, laser altimeters that send pulses of visible light toward Earth's surface have been used for measuring ice sheet mass balance, cloud property information, and topography. ICESat (Ice, Cloud, and land Elevation Satellite) is part of NASA's Earth Observing System, which was launched into near-polar orbit on January 13, 2003, and operated until February 2010 [34]. The sole sensor payload is the Geoscience Laser Altimeter System (GLAS), a next-generation space-based LiDAR that combines strengths of surface LiDAR and dual wavelength cloud and aerosol LiDAR systems. It has three lasers emitting laser pulses at 1,064 and 532 nm wavelengths, capable of producing approximately 70 m diameter laser spots spaced nearly 170 m (560 ft) apart along the spacecraft's ground track [35]. ICESat was designed to study ice-covered regions with specific coverage over the Greenland and Antarctic ice sheets. ICESat laser imaging has contributed to understanding how changes in Earth's atmosphere and climate affect polar ice masses and global sea level.

**Fig. 9.8** ICESat, CryoSat and PIOMAS sea ice thickness measurements in the Arctic. **a** shows the 2003–2007 average ICESat ice thickness for October/November and **b** the 2004–2008 average for February/March. **c–f** are measurements based on CryoSat data, for **c** October/November 2010, **d** February/March 2011, **e** October/November 2011 and **f** February/March 2012. The final two figures are based on PIOMAS measurements for **g** October/ November 2011 and **h** February/March 2012. (Courtesy of ESA) [33]

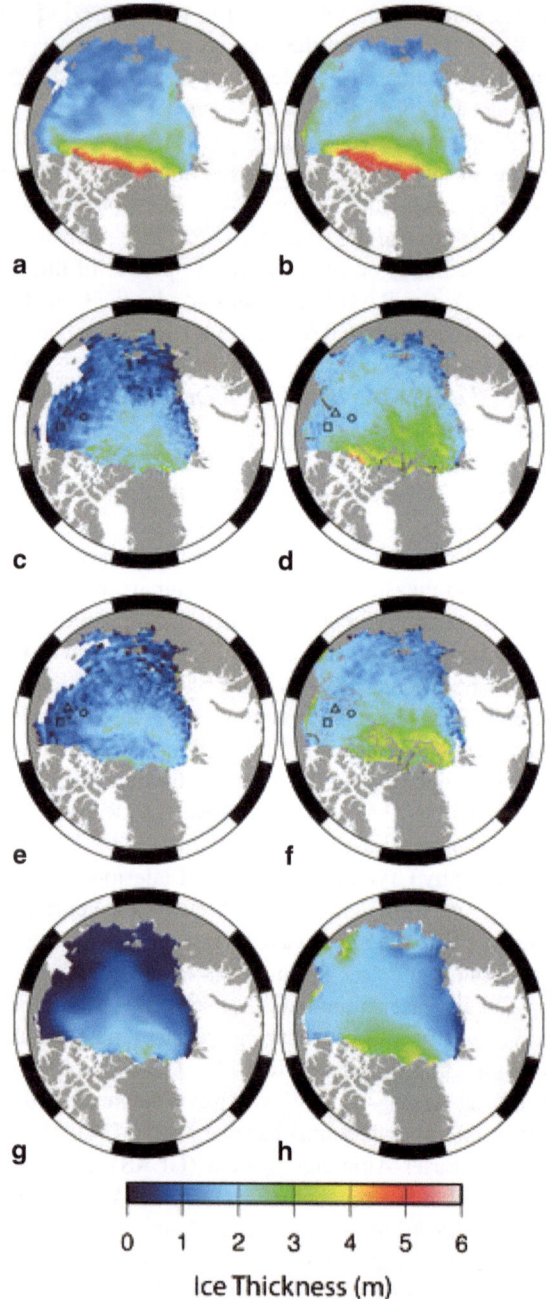

Ice Thickness (m)

# Meeting Long-term Climate Monitoring Requirements

Global climate change has already had significant effects on the environment, which are observable from satellite platforms in space. Effects that scientists had predicted would result from global climate change are now occurring, including loss of sea ice, accelerated sea level rise, more intense heat waves, and more extreme weather events, such as droughts and flooding, as well as more destructive violent storms. According to the IPCC, the extent of climate change effects on individual regions will vary over time, along with the ability of different societal and environmental systems to mitigate or adapt to change. Assessment of Earth observation requirements for adaptation to climate variability and climate change is needed, including considering impacts on sectors related to agriculture, water resources, health, and energy. Identification of observational requirements for adaptation can be better achieved by reviewing the GCOS ECVs and addressing gaps and deficiencies in climate-related observation and monitoring.

As previously discussed, GCOS has maintained an important role for ensuring that data needs are met for climate system monitoring. This includes observing ECVs and assessing the impacts of climate variability and change and implications for national economic development. It fosters research leading to improved understanding, modeling, and prediction of the climate system. The primary goal of the GCOS is to provide continuous, reliable, comprehensive data, and information on the status of the global climate system, building on the World Weather Watch, Global Observing System (GOS), the Global Atmosphere Watch (GAW), the Global Ocean Observing System (GOOS), and the Global Terrestrial Observing System (GTOS). GCOS essentially forms the climate (observing) component of the future GEOSS [36].

Despite recent advances in Earth observation technology, the task of defining observing requirements is still evolving, especially when user needs are considered. A major challenge in long-term monitoring of the climate system is developing and maintaining observing programs, particularly in remote regions of the world. Moreover, many developing countries and economies in transition are greatly challenged when implementing and sustaining even the most basic observing systems. Many countries lack the infrastructural, technical, human and institutional capacities to provide high-quality climate services (Figs. 9.9 and 9.10). As a result, the spatial coverage of *in situ* climate observing networks have been deteriorating globally since the 1990s, and have been rendered inadequate for monitoring regional and local climate change or outputting parameters necessary for climate models. This results in the need for more capable space-based meteorological systems.

The Global Framework for Climate Services (GFCS) is a collective of governments and organizations that produce and use climate information and services [38]. Established at the World Climate Conference-3 in 2009, GFCS seeks to "enable better management of the risks of climate variability and change and adaptation to climate change, through the development and incorporation of science-based climate information. It also seeks to improve climate prediction and to improve planning,

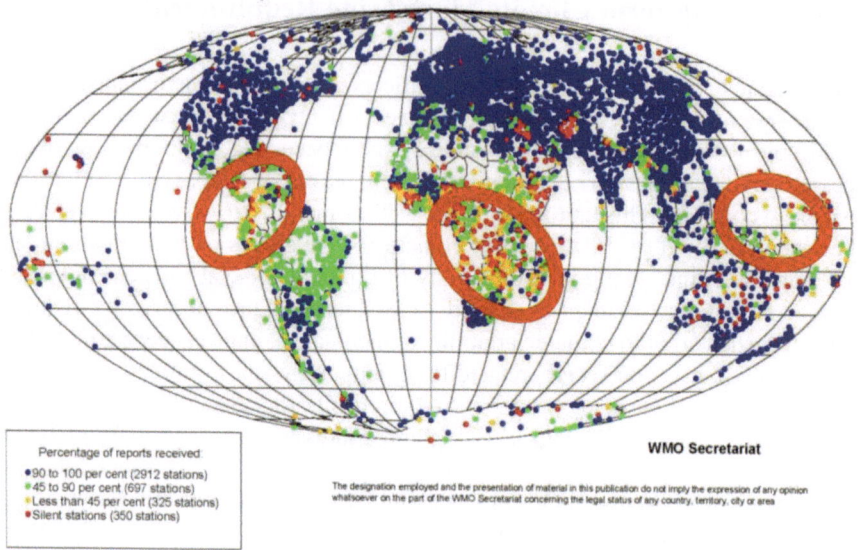

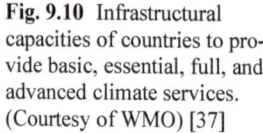

**Fig. 9.9** Annual Global Monitoring 1-15/10/2008 SYNOP reports made at 00, 06, 12, and 18 UTC at Regional Basic Synoptic Network stations. (Courtesy of WMO) [37]

**Fig. 9.10** Infrastructural capacities of countries to provide basic, essential, full, and advanced climate services. (Courtesy of WMO) [37]

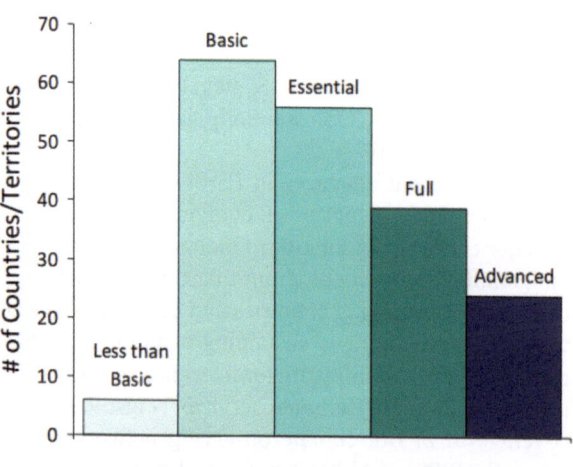

policy, and practice with regard to all types of climate services on the global, regional, and national scale [39]." Finally, it aims to establish a robust and adequately resourced global framework to improve the quality and quantity of climate services worldwide, particularly in developing countries. The GFCS effectively includes GCOS as an observations and monitoring component, along with the World Climate Research Program (WCRP) and a new World Climate Services Program (WCSP).

**Fig. 9.11** Components of the
GFCS framework. (Courtesy
of GFCS) [41]

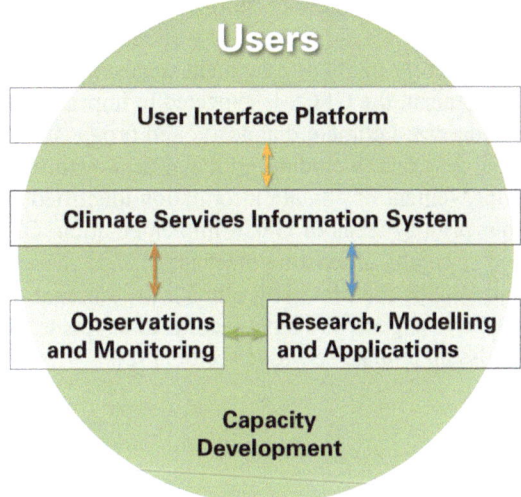

The goals of the GFCS are [40]:

1. reducing the vulnerability of society to climate-related hazards through better provision of climate services.
2. advancing the key global development goals through better provision of climate services.
3. mainstreaming the use of climate information in decision making. Promoting better uptake, understanding and awareness of the need for climate information and climate services, and demonstrating the value of the services in socio-economic, safety, and sustainability terms.
4. strengthening the engagement of providers and users of climate services. Building relationships between providers and users of climate services at both the technical and decision-making levels.
5. maximizing the utility of existing climate service infrastructure. Improving coordination, and strengthening and building this infrastructure where needed.

As shown in Fig. 9.11, the GFCS framework involves five components. First, the 'User Interface Platform' will provide ways for climate service users and providers to interact and improve the effectiveness of the framework and its climate services. Second, the 'Climate Services Information System' will produce and distribute climate data and information according to the needs of users and to agreed standards. Third, 'Observations and Monitoring' will involve developing agreements and standards for generating necessary climate data. Fourth, 'Research, Modeling and Prediction' will harness science capabilities and results to meet the needs of climate services. Finally, 'Capacity Development' will support the systematic development of the institutions, infrastructure and human resources needed for effective climate services. GFCS operations will be divided into global, regional, and local levels in order to tailor information to user requirements for accessing data and knowledge products. Currently, four short-term priority areas of the GFCS have been identified as agriculture and food security, disaster risk reduction, health, and water.

Currently in the development stages of its implementation plan and governance mechanism, the GFCS is expected to improve the availability of regularly updated standardized climate diagnostic and prognostic information. It will also focus climate research, including pilot and demonstration projects, on delivering sustained improvement of climate information identified as feasible and most needed in the four priority areas of GFCS implementation. GFCS will build upon the resources and strengths of existing mechanisms and institutions, including the GCOS, GEO, and WCRP for providing climate information and services. By the end of 2021, GFCS is expected to have facilitated access to improved climate services globally across all climate sensitive sectors [42].

## Summary

With various institutions and coordinating bodies providing climate services, including the GFCS, such efforts will progress to a better understanding of climate system variability and change. This may provide more predictive knowledge when developing adaptation and mitigation strategies at regional and national scales. Such strategies will assist planning for climate change impacts on social and economic sectors, including food security, energy, transport, environment, health, and water resources. Numerical prediction models for simulating and assessing the climate system will continue to rely on observations of components of the Earth system, including the atmosphere, oceans, land, and cryosphere. Meteorological satellites will continue to play an increasingly important role in providing a continuous stream of measurements and data for numerical analysis, and for the development of science-based climate services.

## References

1.  IPCC AR4 WG1.: IPCC Fourth Assessment Report: Climate Change 2007: The Physical Science Basis. Solomon, S., Qin, D., Manning, M., Chen, Z., Marquis, M., Averyt, K.B., Tignor, M., and Miller, H.L., ed. Contribution of Working Group I to the Fourth Assessment Report of the Intergovernmental Panel on Climate Change, Cambridge University Press (2007)
2.  IPCC AR4 WG2.: IPCC Fourth Assessment Report: Climate Change 2007: Impacts, Adaptation and Vulnerability. Parry, M.L., Canziani, O.F., Palutikof, J.P., van der Linden, P.J., and Hanson, C.E., ed. Contribution of Working Group II to the Fourth Assessment Report of the Intergovernmental Panel on Climate Change, Cambridge University Press (2007)
3.  WMO.: http://www.wmo.int/pages/prog/gcos/index.php?name=EssentialClimateVariables. Accessed 12 Aug 2013
4.  CEOS.: http://www.eohandbook.com/eohb2011/climate_variables.html. Accessed 12 Aug 2013
5.  WMO.: Implementation Plan for the Global Observing System for Climate in Support of the UNFCCC (2004)

6. Li, J., Schmidt, C.C., Nelson, J.P., Schmit, T.J., Menzel, W.P.: Estimation of total atmospheric ozone from GOES sounder radiances with high temporal resolution. Am. Meteorol. Soc., **18**, 157–168 (2001)

7. Orsolini, Y.J., Karcher, F.: Total ozone imaging over North America with GOES-8 infrared measurements. Q. J. Roy. Meteor. Soc. , **126**(565), 1557–1561 (2000)

8. Menzel, W.P., Holt, F.C., Schmit, T.J., Aune, R.M., Schreiner, A.J., Wade, G.S., Gray, D.G.: Application of GOES-8/9 soundings to weather forecasting and nowcasting. B. Am. Meteorol., **79**, 2059–2077 (1998)

9. Menzel, W.P., and Purdom, J.F.W.: Introducing GOES-I: The first of a new generation of geostationary operational environmental satellites. B. Am. Meteorol., **75**, 757–781 (1994)

10. CIMSS (Cooperative Institute for Meteorological Satellite Studies, University of Wisconsin-Madison).: http://cimss.ssec.wisc.edu/goes/misc/990929.html. Accessed 12 Aug 2013

11. NESDIS.: http://www.orbit.nesdis.noaa.gov/smcd/spb/ozone/. Accessed 12 Aug 2013

12. NASA.: http://disc.sci.gsfc.nasa.gov/acdisc/TOMS. Accessed 12 Aug 2013

13. NASA.: http://ozoneaq.gsfc.nasa.gov/omps/. Accessed 12 Aug 2013

14. NASA.: http://npp.gsfc.nasa.gov/omps.html. Accessed 12 Aug 2013

15. Kidd, C.: Satellite rainfall climatology: A review. **International Journal of Climatology**, **21**, 1041–1066 (2001)

16. Druen, B., Heinemann, G.: Rain rate estimation from a synergetic use of SSM/I, AVHRR and meso-scale numerical model data. Meteorol. Atmos. Phys. , **66**, 65–85 (1998)

17. Scofield, R.A.: The NESDIS operational convective precipitation estimation technique. Mon. Weather. Rev. , **115**, 1773–1792 (1987)

18. Arkin, P., Meisner, B.N.: The relationship between large-scale convective rainfall and cold cloud cover over the Western hemisphere during 1982–84. Montly Weather Review, 115, 51–74 (1987)

19. GEWEX.: http://www.gewex.org/gpcp.html. Accessed 12 Aug 2013

20. NOAA.: http://www.cpc.ncep.noaa.gov/products/global_precip/html/wpage.cmap.html. Accessed 12 Aug 2013

21. NOAA.: http://www.cpc.ncep.noaa.gov/products/janowiak/cmorph_description.html. Accessed 12 Aug 2013

22. NASA.: http://trmm.gsfc.nasa.gov/. Accessed 12 Aug 2013

23. NASA.: http://trmm.gsfc.nasa.gov/publications_dir/rumbia_june-july_2013.html. Accessed 12 Aug 2013

24. NASA.: http://www.nasa.gov/mission_pages/GPM/main/index.html. Accessed 12 Aug 2013

25. NOAA.: http://www.elnino.noaa.gov/. Accessed 12 Aug 2013

26. Ferraro, R., Weng, F., Grody, N., Basist, A.: An eight year (1987–1994) time series of rainfall, clouds, water vapor, snow-cover, and sea-ice derived from SSM/I measurements. B. Am. Meteorol., **77**(5), 891–905 (1996)

27. Curtis, S.: Evolution of El Niño – Precipitation relationships from satellites and gauges. Journal of Geographical Research, **108**, 1–8 (2003)

28. NASA.: http://science1.nasa.gov/earth-science/oceanography/physical-ocean/ocean-surface-topography/. Accessed 12 Aug 2013

29. NSIDC.: http://nsidc.org/cryosphere/seaice/. Accessed 12 Aug 2013

30. Pielke, R.A., Liston, G.E., Chapman, W.L., Robinson, D.A.: Actual and insolation-weighted Northern Hemisphere snow cover and sea-ice between 1973–2002. Clim. Dynamics, **22**, 591–595 (2004)

31. Canadian Space Agency.: http://www.asc-csa.gc.ca/eng/satellites/radarsat2/. Accessed 12 Aug 2013

32. eoPortal.: https://directory.eoportal.org/web/eoportal/satellite-missions/c-missions/cryosat. Accessed 12 Aug 2013

33. ESA.: http://www.esa.int/Our_Activities/Observing_the_Earth/CryoSat/CryoSat_reveals_major_loss_of_Arctic_sea_ice. Accessed 12 Aug 2013

34. NASA.: http://icesat.gsfc.nasa.gov/icesat/. Accessed 12 Aug 2013

35. NASA.: http://earthobservatory.nasa.gov/Features/ICESat/. Accessed 12 Aug 2013

36. Houghton, J., Townshend, J., Dawson, K., Mason, P., Zillman, J., Simmons, A.: The GCOS at 20 years: the origin, achievement and future development of the Global Climate Observing System. Weather, **67**(9), 227–235 (2012)
37. WMO Presentation.: The Global Framework for Climate Services (GFCS) by J. Lengoasa, Deputy Secretary-General. http://wcrp-climate.org/JSC33/presentations/GFCS.pdf. Accessed 12 Aug 2013
38. GFCS.: http://www.gfcs-climate.org. Accessed 12 Aug 2013
39. World Climate Conference-3.: (August31 to September 4, 2009) http://www.wmo.int/wcc3/page_en.php. Accessed 12 Aug 2013
40. GFCS.: http://www.gfcs-climate.org/content/about-gfcs. Accessed 12 Aug 2013
41. GFCS.: http://www.gfcs-climate.org/components. Accessed 12 Aug 2013
42. WMO.: Draft Global Framework for Climate Services. Concept Note, Ver 3.4, March 11, 2009 (2009)

# Chapter 10
# Top Ten Things to Know About Meteorological Satellites

> *"Everyone talks about the weather, but nobody does anything about it."*
>
> —Mark Twain (1835–1910)

Mark Twain did indeed have plenty to say about the weather. He also famously said, "it is best to read the weather forecast before we pray for rain." He was likely aware that something could be done about forecasting the weather, namely by observing and understanding the weather, rather than attempting to directly influence or control it.

Weather forecasting today still takes the same approach of gathering weather observations (by satellites, ground measurements, and weather balloon soundings) to forecast weather at a particular location, nowadays preferably in real-time. Networks of weather monitoring stations and space-based satellites are currently tracking weather systems and the physics and dynamics of the atmosphere. This is done by feeding observations into sophisticated Numerical Weather Prediction (NWP) models. Now-casting (or updated weather forecasting within a 6-hour timeframe) relies on the latest radar, satellite, and observational data to attain accurate and small-scale forecasts.

International cooperation for meteorological data exchange is paramount, especially since there are limitations to the density of observing networks. Not all nations can afford such systems for forecasting the weather and potential hazards. Furthermore, weather systems are not constrained to national borders, and sparsely populated or remote regions may not have adequate observation capabilities.

This has led to the establishment of the World Meteorological Organization (WMO) in 1950 as a specialized United Nations agency for facilitating international cooperation in meteorology and promoting research. Indeed, the WMO is considered to be a success story, both in terms of meeting its primary goal of enabling weather services around the world to better serve their users, and also in terms of providing a model for successful international collaboration on a global scale.

This book has attempted to provide a comprehensive overview of meteorological satellites and their Earth applications. It is a multi-disciplinary reference that introduces space meteorology technologies and national systems, as well as international efforts to tackle climate change. The important take-away lessons from this book are summarized below as a list of "top ten things to know about meteorological satellites."

S.-Y. Tan, *Meteorological Satellite Systems,* SpringerBriefs in Space Development, DOI 10.1007/978-1-4614-9420-1_10, © The author 2014

1. **Both polar-orbiting and geosynchronous satellites are now key complements to traditional methods of weather forecasting.** Polar-orbiting satellites operate at a lower position (about 840 km above Earth), passing over a different swath of Earth in each pass. Geostationary satellites are placed into orbit at about 36,000 km above Earth's equator, fixed in relation to Earth and providing a constant view of the geographic area below. These satellites collectively provide a complete global weather monitoring system. Orbital characteristics for observing Earth help to determine the types of applications for which a weather satellite is best suited. Modern weather satellites and sensors have evolved from a long history of collecting meteorological observations from a vast array of early instrumentation, including weather balloons, radiosondes, kites, and weather reconnaissance aircraft. Although precise in measurements, historical weather instruments and platforms were limited in providing regional or global coverage of observations. Real-time and synoptic monitoring of large areas of Earth by satellites has allowed improved meteorological data collection. Earth-observing satellites offer advantages of being able to collect meteorological data at synoptic scales, in remote locations, and providing real-time monitoring capabilities.

2. **Meteorological satellite technologies have evolved to allow for higher resolution sensing, merging of data from different types of satellites in different orbits, and combining data from satellites of different countries to strengthen global weather forecasting capabilities.** These technologies have evolved from television cameras in the early 1960s to multi-channel high-resolution radiometers in the 1970s. Atmospheric sounding systems enable vertical profiles of temperature, pressure, water vapor, and atmospheric gases to be collected, which further enhances understanding of cloud formation processes and weather forecasting capabilities. Integration of multi-sensor data is an active area of research and international satellite data exchange is required for enhancing weather prediction and advanced warnings for protecting people and property.

3. **The U.S. meteorological satellite program has been developed, launched, and operated by the tri-agency partnership and coordination of NASA, NOAA, and the Department of Defense.** Satellites for civilian and military usage have developed largely as separate systems, although there have been proposals to merge the system to save on development and operational costs. The TIROS (Television Infrared Observation Satellite) and Nimbus spacecrafts were launched as some of the first polar-orbiting meteorological satellites in the 1960s, while experimental ATS (Applications Technology Satellites) were launched as geosynchronous weather satellites. The next-generation polar-orbiting satellites include the Joint Polar Satellite System (JPSS) and Suomi National Polar-orbiting Partnership (SNPP), formally launched in 2011. GEOS-R is the latest geostationary satellite effort expected to launch in 2015, providing higher spectral, spatial, and temporal information, real-time mapping of lightning activity, and improved severe weather forecasting and solar monitoring capabilities. There has been increased cooperation and integration of the U.S. tri-agency meteorological satellite partnership that appears to be paying dividends in increased coverage at lower cost.

4. **The European meteorological satellite system maintains both geostationary and polar-orbiting satellite systems, which are launched by the European Space Agency (ESA) and operated by the European Organization for the Exploitation of Meteorological Satellites (EUMETSAT).** The Meteosat system was initiated in 1972 and became the first European meteorological program in geostationary orbit. This system's operations were transferred from ESA to EUMETSAT in 1987. EUMETSAT has since launched and operated geostationary meteorological satellites under the Meteosat Operational Program (MOP), Meteosat Transition Program (MTP), and Meteosat Second Generation (MSG) program. Polar-orbiting systems are also maintained to provide more detailed measurements of the atmosphere and improved synoptic coverage. This will be accomplished with the EUMETSAT Polar System (EPS) that consists of the Meteorological Operational Satellite Program of Europe (MetOp) and a follow-on Polar System Second Generation (EPS-SG) program planned for 2020.

5. **Meteorological satellite systems have been developed by other countries, including Russia, China, Japan, and India, which collectively provide important input into the World Weather Watch for global weather forecasting.** Significant advances in spacecraft design, sensor technologies, data exchange, and operations have been attained by national investments in meteorological satellite-sensing capabilities. Some nations have developed both geostationary and polar-orbiting meteorological satellite systems to meet growing demands for weather information. For example, China is investing heavily into its next-generation of FengYun meteorological satellites and India has improved payloads in its INSAT geostationary systems. Significant upgrades in Russian and Japanese systems have likewise been achieved.

6. **Deployment of new meteorological satellite systems creates an increasing need for improved international coordination and cooperation for data sharing, distribution, and global weather forecasting and severe weather warnings.** This is necessary to guarantee timely data access for maximizing societal benefits of meteorological satellite systems. These systems collectively form part of the Global Observing Systems (GOS), which contributes to the development of the Global Earth Observation System of Systems (GEOSS) coordinated by the Group on Earth Observations (GEO). The deployment of meteorological satellites and coordinated sharing of data from these networks are perhaps the most important new development in weather forecasting and monitoring of climate change.

7. **Worldwide weather data collection networks allow effective sharing of meteorological data on a global scale, which is essential for predicting weather and providing early severe weather warnings for saving lives and property.** Founded in 1873, the World Meteorological Organization (WMO) is an intergovernmental organization and specialized agency of the United Nations for meteorology (weather and climate), operational hydrology, and related geophysical sciences. The WMO has played a crucial role in improving global meteorological activities. As of January 2013 the WMO included 191 member states and territories. This global U.N. agency has provided a unique and relatively

successful mechanism for timely and unrestricted exchange of data, information, and products among its members to facilitate real- or near-real time forecasts and early warnings. The WMO World Weather Watch (WWW) program, drafted in 1963, has played a key role in coordinating the collection and distribution of meteorological data for weather forecasts. This includes the space-based Global Observing System (GOS), which consists of various observation platforms and components for weather data collection.

8. **The Group on Earth Observations (GEO) is coordinating efforts to build a comprehensive, coordinated, and sustained Earth observation system, called the Global Earth Observation System of Systems (GEOSS).** The effort to develop GEOSS was launched in response to the 2002 World Summit on Sustainable Development. It is based on a 10-Year Implementation Plan for 2005 to 2015. The goal of GEOSS is to provide comprehensive coordinated Earth observations from national, regional, and international systems and instruments, working towards full and open data sharing and improved data and information access. This will necessitate common standards for data architecture and sharing for interlinking Earth observation systems. Collected data will focus on nine "Societal Benefit Areas" identified as being critically important to people and society, including disasters, health, energy, climate, and water. GEOSS is intended to serve a wide range of users as a decision-support tool, including public and private sectors, resource managers, planners, emergency responders, and scientists. This initiative will feed the growing demand for Earth observation data and information, as well as supporting and engaging users in developing countries. It will also provide many benefits, but perhaps the greatest benefit will be in monitoring and understanding trends related to climate change.

9. **Future development of meteorological satellite programs is moving towards enhancing now-casting abilities and producing higher resolution products for supporting weather monitoring, severe weather warning, and disaster mitigation.** Meteorological satellites have rapidly evolved to improve observation of Earth's atmosphere and to provide more accurate weather forecasts. Now-casting provides detailed descriptions of weather conditions in real-time or extrapolated for a period of 0 to 6 hours ahead. This is achieved by integrating observations from a variety of sources in addition to satellite data, including radar, weather station, radiosonde, and other observational information. This collected data is essential for observing climate variability and predicting severe weather conditions, which is important for a range of applications in aviation, navigation, agriculture, forestry, hydrology, and other economic sectors. Now-casting is a powerful location-specific forecasting tool, which is further enhanced by new satellite technologies and sensor instrumentation, such as multispectral and hyperspectral imagers, lightning mappers, solar monitoring instruments, and direct broadcast for real-time delivery of weather data products.

10. **Meteorological satellites are used to monitor long-term climate change, including the ozone hole, precipitation and cloud patterns, ocean dynamics, and sea ice.** The Global Climate Observing System (GCOS) is an internationally coordinated observing system established in 1992 for providing

continuous, worldwide, reliable, comprehensive data and information on the state and behavior of the global climate system. It has defined and regularly updated climate requirements for global observing systems, providing observations of Essential Climate Variables (ECVs). International exchange of both current and historical observations of Earth from satellite-based monitoring is required for making longer-term climate predictions and for better understanding our changing climate.

This book demonstrates the valuable role that meteorological satellites play in improving human safety and security. In fact, weather data is often considered to be a 'global public good.' This has led to international collaboration and data sharing and in particular has given rise to initiatives such as the Group on Earth Observations (GEO). Indeed international cooperation remains a cornerstone of international satellite meteorology today, with increasing efforts for building capacity in developing countries and remote regions. There is also a growing need for data integration of various satellite sensors and traditional-collected weather data, which helps improve weather forecasts and severe weather predicting.

Substantial changes are expected to be made to further enhance operational meteorological systems and sensor technologies. The growing role of other satellite providers, such as China, Russia, Japan, and India, are likely to ensure that geostationary satellite coverage will be enhanced with a greater number of weather satellites in orbit, enhanced payloads being developed, and a growth in polar-orbiting systems to provide synoptic coverage at regional and global scales. The rise of new players will likely be a gain for all.

The rapid evolution of meteorological satellite technologies provides improved observation and understanding of Earth's environment, especially of hazards and extreme weather events. Such economic and social impacts and benefits of weather forecasts and atmospheric sciences are difficult to assess and quantify properly. With human adaptation and mitigation to the impacts of climate change in focus, there will be an ever-increasing and urgent need to improve our understanding of global climate and weather patterns. This is indeed an exciting time for observing the evolving future technologies of next-generation meteorological satellite systems. We hope these new systems can help us answer the old-age question of what the weather will be like tomorrow.